AF627374

ISSN 0044-2267

ZEITSCHRIFT FÜR ANGEWANDTE MATHEMATIK UND MECHANIK

INGENIEURWISSENSCHAFTLICHE FORSCHUNGSARBEITEN

UNTER MITWIRKUNG VON E. BECKER · H. BECKERT · L. BERG · L. BITTNER · L. COLLATZ W. FISZDON · H. GÖRTLER · H. GAJEWSKI · J. HEINHOLD · H. HEINRICH · J. HULT A. JU. ISCHLINSKI · R. KLÖTZLER · P. H. MÜLLER · H. NEUBER · W. OLSZAK · K. OSWATITSCH A. SAWCZUK · L. SCHMETTERER · J. W. SCHMIDT · H. SCHUBERT · G. G. TSCHORNY H. UNGER · F. WEIDENHAMMER UND F. ZIEGLER
HERAUSGEGEBEN VON G. SCHMIDT, BERLIN

CHEFREDAKTEUR: G. SCHMIDT REDAKTEURE: W. HEINRICH, H. WEINERT

BAND 60 1980 HEFT 2

INHALT

AKADEMIE-VERLAG · BERLIN

ZAMM Bd. 60 Nr. 2 S. 65—128 Berlin, Februar 1980

EVP 18,— M 34115

INHALT

Hauptaufsätze

Kleine Mitteilungen

Wir bitten, **Manuskriptsendungen** zweifach (Original und eine Kopie, sprachlich einwandfrei, Formeln mit Maschine oder in Druckschrift geschrieben) an folgende Anschrift zu richten:

Zeitschrift für Angewandte Mathematik und Mechanik z. Hd. Herrn Prof. Dr. Günter Schmidt
Zentralinstitut für Mathematik und Mechanik an der Akademie der Wissenschaften der DDR
DDR-1080 Berlin, Mohrenstraße 39.

Zu den Arbeiten, die als Hauptaufsätze bestimmt sind, ist auf gesondertem Blatt eine Zusammenfassung von 5 bis 10 Zeilen in englischer und (möglichst) deutscher und russischer Sprache beizufügen. Ausführliche Hinweise für die Autoren, um deren strikte Berücksichtigung gebeten wird, finden sich im Anschluß an das Inhaltsverzeichnis des Jahrganges **59** (1979).

Die Autoren erhalten von den Hauptaufsätzen 75, von den Kleinen Mitteilungen 25 Sonderdrucke ohne Berechnung, darüber hinaus weitere Sonderdrucke gegen Berechnung.

Der Verlag behält sich für alle Beiträge das Recht der Vervielfältigung und Übersetzung vor.

Bestellungen sind zu richten

— **in der DDR** an den Postzeitungsvertrieb, an eine Buchhandlung oder an den Akademie-Verlag, DDR-1080 Berlin, Leipziger Straße 3—4
— **im sozialistischen Ausland** an eine Buchhandlung für fremdsprachige Literatur oder an den zuständigen Postzeitungsvertrieb
— **in der BRD und Westberlin** an eine Buchhandlung oder an die Auslieferungsstelle KUNST UND WISSEN, Erich Bieber, D-7000 Stuttgart 1, Wilhelmstraße 4—6
— **in Österreich** an den Globus-Buchvertrieb, A-1201 Wien, Höchstädtplatz 3
— **in den übrigen westeuropäischen Ländern** an eine Buchhandlung oder an die Auslieferungsstelle KUNST UND WISSEN, Erich Bieber GmbH, CH-8008 Zürich/Schweiz, Dufourstraße 51
— **im übrigen Ausland** an den Internationalen Buch- und Zeitschriftenhandel; den Buchexport, Volkseigener Außenhandelsbetrieb der Deutschen Demokratischen Republik, DDR-7010 Leipzig, Postfach 160, oder an den AKADEMIE-VERLAG, DDR 1080 Berlin, Leipziger Straße 3—4.

ZEITSCHRIFT FÜR ANGEWANDTE MATHEMATIK UND MECHANIK
Herausgeber und Chefredakteur: Prof. Dr. Günter Schmidt. Redaktion: Dr. Winfried Heinrich, Dr. Horst Weinert, Dipl.-Math. Friedhild Dudel, Helga Rühl, Zentralinstitut für Mathematik und Mechanik der Akademie der Wissenschaften der DDR. Verlag: Akademie-Verlag, DDR-1080 Berlin, Leipziger Straße 3—4; Fernruf 2236221 oder 2236221; Telex-Nr.: 114420; Bank: Staatsbank der DDR, Berlin, Kto.-Nr.: 6836-26-20712. Anschrift der Redaktion: Zentralinstitut für Mathematik und Mechanik der Akademie der Wissenschaften, DDR-1080 Berlin, Mohrenstraße 39; Fernruf: 2000561.

Veröffentlicht unter der Lizenznummer 1282 des Presseamtes beim Vorsitzenden des Ministerrates der Deutschen Demokratischen Republik. Gesamtherstellung: VEB Druckerei „Thomas Müntzer", DDR-5820 Bad Langensalza. Erscheinungsweise: Die Zeitschrift für Angewandte Mathematik und Mechanik erscheint monatlich. Die 12 Hefte eines Jahres einschließlich Tagungsheft bilden einen Band. Bezugspreis je Band 360,— M zuzüglich Versandspesen (Preis für die DDR 216,—M). Bezugspreis je Heft 30,—M (Preis für die DDR 18,—M). Bestellnummer dieses Heftes: 1009/60/2.

© 1980 by Akademie-Verlag Berlin · Printed in the German Democratic Republic.

AN (EDV) 35937

ASMW

ZAMM 60, 65 – 70 (1980)

M. Epstein / P. G. Glockner

On Non-Isothermal Rigid Shell Theory*)

Using the two fundamental laws of Thermodynamics, results are derived for placing restrictions on constitutive equations of rigid shells. This is achieved by dividing the shell into a number of fictitious layers and accommodating an arbitrary temperature field by use of a continuous piecewise linear coldness function across the thickness. Layer entropy densities are introduced in terms of which the entropy inequality is globally valid for each layer.

Unter Benutzung der zwei Fundamentalsätze der Thermodynamik werden Restriktionsbedingungen für die Konstitutivgleichungen von starren Schalen erhalten. Dies wird durch Unterteilung der Schale in eine Anzahl von fiktiven Schichten und durch Anpassung eines willkürlichen Temperaturfeldes erreicht, wobei eine stetige stückweise lineare Kältefunktion über die Dicke benutzt wird. Es werden Schichtentropiedichten eingeführt, mittels derer die Entropieungleichung für jede Schicht erfüllt wird.

Используя две основные теоремы термодинамики, получены условия связи для уравнения состояния жёсткой оболочки. Это достигнуто путём расчленения оболочки на некоторое число фиктивных слоёв и путём приведения в соответствие произвольного поля температур, причём, использовалась непрерывная кусочнолинейная функция холода по толщине. В ведены плотности энтропии слоёв, посредством которых выполнялось неравенство энтропии для каждого слоя.

1. Introduction

Despite the obvious practical relevance of the non-isothermal deformations of plates and shells, the subject has not been extensively dealt with in the literature. The usual way to accomodate temperature gradients in the direction normal to the reference surface is to employ the engineering concept of thermal stresses, e.g. [1], which give rise to equivalent thermal forces and couples. This treatment, although useful in many practical situations, is of limited scope and is not in line with the recent trends of Thermodynamics of Continua.

Complete thermodynamic treatments which are applicable to shell theory have been given in the works of Grot [2], Koh and Weng [3] and Green and Naghdi [4] (see also Krätzig [5]). Both in [2] and [3], the authors deal with the thermodynamics of materials with microstructure which is readily translatable into shell theory language, whereas the subject matter in [4] is directly concerned with our topic.

In their work, Koh and Weng adopt Biot's variational approach [6] thereby avoiding the question of how the entropy inequality should be formulated. In Grot's work, on the other hand, the entropy inequality is neatly obtained by integrating the classical Clausius-Duhem inequality over the volume of a micro-element and defining a single entropy density for the macro-volume. The temperature distribution in each micro-element is approximated by assuming linearity of the coldness field.

An arbitrary temperature distribution is considered in [4] in which the field equations for the shell are obtained by taking moments of all orders with respect to a reference surface. Although this technique is mathematically correct for the transition from three- to two-dimensional equations, it doesn's appear to be entirely satisfactory for the treatment of inequalities. This peculiarity can perhaps best be understood through a simple example. The linear function

$$f(z) = 2z - 1$$

in the interval [0, 1] has all of its moments with respect to $z = 0$ nonnegative, although the function itself becomes negative in the finite subinterval [0, 1/2]. In the particular case of a linear function it is not difficult to find the conditions on the zero-th and first order moments which would guarantee the non-negativeness of the function on a given finite interval. For second, third and fourth order polynomials, however, the appropriate conditions become more complicated (and non-linear), and for higher order polynomials they would be, in general, impossible to obtain in an explicit way, as these conditions are connected with the location of the roots of the polynomial.

In the present paper an attempt is made to accommodate an arbitrary temperature field by approximating the coldness by means of a continuous piecewise linear function across the thickness rather than by a polynomial expansion. The shell is thus conceived as a multilayered structure in which for each fictitious layer a treatment comparable to that of [2] may be justified. Thus the entropy inequality is globally valid for each layer in terms of a layer entropy density. In the limit, as the number of layers is increased, the local three-dimensional entropy inequality will be satisfied. The treatment is thus reminiscent of techniques used in finite element analysis. If in the formulation presented below, not only the coldness but also the displacement fields are approximated by continuous piecewise linear functions, the analysis becomes a generalization of [7], in which thermal effects were ignored. It is conceivable, however, that the displacement fields could.be assumed to be only linear throughout the thickness, as in classical monocoque shell theory, while the temperature is conceived in terms of layers. In order to exhibit the features of our approach in as simple a way as possible, only the case of a rigid heat-conducting shell is treated here. The generalization for deformable shells will be the subject of a further paper.

*) The results presented here were obtained in the course of research sponsered by the National Research Council of Canada, Grant Nos. A 4662 and A 2736.

2. On Layer Coordinates and the Representation of Continuous piece-wise Linear Functions

Let the shell be subdivided into N layers and denote by h_I ($I = 1, \ldots, N$) the thickness of the I-th layer measured in the direction normal to the reference surface at a point P.

A set of non-dimensional layer coordinates, z^I ($I = 1, \ldots, N$), is then defined at P in the following manner [7, 8]:

a) Starting at P and following the normal to the reference surface, the first point of layer I encountered is called the origin, $O_I(P)$, of that layer corresponding to P. Thus the origin of any layer is at an interface between layers, except for the layer containing the reference surface, for which the origin is P itself.

b) At layer I, the value of z^J is calculated as follows:
(i) for $J = I$: $z^J h_I$ measures length along the normal, starting from zero at the origin $O_I(P)$ and increasing in the positive sense of the normal;
(ii) for $J \neq I$: z^J is a constant equal to the last value attained by z^J in layer J when travelling from P towards layer I, or zero if layer J is not traversed in that trajectory.

Let $f(z)$ be a continuous piecewise linear function of the normal thickness coordinate, z. Then if the "kinks" of $f(z)$ are at the interfaces between layers, one can write

$$f(z) = f_0 + f_I z^I , \tag{1}$$

where f_0 and f_I are constants and where the summation convention applies.

The attractive feature of Eq. (1) is that it has the formal appearance of a linear function. Now, a linear function defined on a finite interval is completely determined by its zero-th and first order moments. In the case of piecewise linear function, Eq. (1) suggests to define the zero-th moment

$$F^0 = \int f(z)\, \mathrm{d}z \tag{2}$$

and the N first order moments

$$F^I = \int f(z)\, z^I\, \mathrm{d}z , \tag{3}$$

where z denotes the usual normal coordinate and where the integrals extend over the thickness and satisfy, therefore

$$\int [\;]\, \mathrm{d}z = \sum_{I=1}^{N} h_I \int [\;]\, \mathrm{d}z^I , \tag{4}$$

in which the right-hand-side integrals extend over the unit interval associated with the corresponding layer.

Substituting Eq. (1) into (2) and (3) yields

$$F^0 = f_0 S^0 + f_I S^I , \qquad F^I = f_0 S^I + f_J S^{JI} \tag{5), (6}$$

where

$$S^0 = \int \mathrm{d}z = \sum_{I=1}^{N} h_I , \qquad S^I = \int z^I\, \mathrm{d}z , \qquad S^{JI} = \int z^J z^I\, \mathrm{d}z . \tag{7), (8), (9}$$

Eqs. (5) and (6) can be regarded as a linear system connecting F^0 and F^I with f_0 and f_J by means of the symmetric coefficient matrix

$$[S] = \begin{bmatrix} S^0 & S^J \\ S^I & S^{IJ} \end{bmatrix} \tag{10}$$

so that, barring the singularity of S, knowledge of the $N + 1$ moments, F^0, F^I, is equivalent to knowledge of the function itself, in complete analogy with a linear function.

At this point it appears that the notation could be made even more compact by allowing indices to run from 0 to N. Such indices will be denoted by capital Greek letters. For consistency, one then introduces a "zero-th layer coordinate", z^0, which is equal to 1 identically. Eqs. (1)–(9) are thus encompassed in the formulas

$$f(z) = f_\Lambda z^\Lambda \tag{11}$$

$$F^\Lambda = \int f(z)\, z^\Lambda\, \mathrm{d}z , \qquad F^\Lambda = f_\Delta S^{\Lambda\Delta} \tag{12), (13}$$

and

$$S^{\Lambda\Delta} = \int z^\Lambda z^\Delta\, \mathrm{d}z . \tag{14}$$

An alternative useful way of representing a piece-wise linear function is through the $N + 1$ values it attains at the interfaces (including the extreme facings). Denoting by $\hat{f}_\Lambda$ those values, where $\Lambda = 0$ corresponds to the "inner" face, $\Lambda = 1$ to the interface between the first and second layers, ..., and $\Lambda = N$ to the "outer" face, one can write

$$\hat{f}_\Delta = f_\Lambda Z^\Lambda_\Delta . \tag{15}$$

where Z^{Λ}_{Δ} is the value of the Λ-th layer coordinate on the Δ-th interface. For the sake of clarity we can write Eq. (15) explicitly in matrix form as

$$\begin{Bmatrix} \hat{f}_0 \\ \hat{f}_1 \\ \hat{f}_2 \\ \vdots \\ \hat{f}_k \\ \vdots \\ \hat{f}_{N-1} \\ \hat{f}_N \end{Bmatrix} = \begin{bmatrix} 1 & -1 & -1 & \dots & \alpha & \dots & 0 & 0 \\ 1 & 0 & -1 & \dots & \alpha & \dots & 0 & 0 \\ 1 & 0 & 0 & & \alpha & & 0 & 0 \\ \vdots & \vdots & \vdots & & (\alpha) & & & \\ 1 & 0 & 0 & \dots & \beta & \dots & 0 & 0 \\ \vdots & \vdots & \vdots & & \vdots & & & \\ 1 & 0 & 0 & \dots & \beta & \dots & 1 & 0 \\ 1 & 0 & 0 & \dots & \beta & \dots & 1 & 1 \end{bmatrix} \begin{Bmatrix} f_0 \\ f_1 \\ f_2 \\ \vdots \\ f_k \\ \vdots \\ f_{N-1} \\ f_N \end{Bmatrix} \tag{16}$$

where K is the layer containing the reference surface and α and β are, respectively, the minimum and maximum values attained by z^K. It is easy to verify that the determinant of this matrix does not vanish (its absolute value is always $\beta - \alpha = 1$). Combining Eqs. (13) and (16) yields

$$F^{\Lambda} = A^{\Lambda\Delta}\hat{f}_{\Delta}\,, \tag{17}$$

where

$$A^{\Lambda\Delta} = S^{\Lambda\Gamma}(Z^{-1})^{\Delta}_{\Gamma}\,. \tag{18}$$

Integrals over one layer of the product of two piecewise linear functions will appear below. Introducing the notation

$$C^{\Delta\Gamma\Lambda} = C^{\Gamma\Delta\Lambda} = \int z^{\Gamma}z^{\Delta}\,\mathrm{d}z^{\Lambda} \tag{19}$$

these integrals are expressible as

$$\int f(z)\,g(z)\,\mathrm{d}z^{\Lambda} = f_{\Gamma}g_{\Delta}C^{\Gamma\Delta\Lambda}\,. \tag{20}$$

The particular case in which $g(z) = 1$ yields

$$\int f(z)\,\mathrm{d}z^{\Lambda} = f_{\Gamma}C^{\Gamma 0\Lambda} \tag{21}$$

or, alternatively,

$$\int f(z)\,\mathrm{d}z^{\Lambda} = \tfrac{1}{2}\,(\hat{f}_{\Lambda-1} + \hat{f}_{\Lambda})\,. \tag{22}$$

Note also that

$$C^{\Gamma\Delta 0} = 0\,. \tag{23}$$

Sometimes, instead of the coefficients defined in Eqs. (14) and (19), similar integrals weighted through a known integrable function, w, will be needed, namely

$$S^{\Lambda\Delta}_{(w)} = \int wz^{\Lambda}z^{\Delta}\,\mathrm{d}z \tag{24}$$

and

$$C^{\Delta\Gamma\Lambda}_{(w)} = C^{\Gamma\Delta\Lambda}_{(w)} = \int wz^{\Gamma}z^{\Delta}\,\mathrm{d}z^{\Lambda}\,. \tag{25}$$

Note that discontinuous integrable weighting functions are admissible.

3. Energy Balance

The first law of Thermodynamics for a rigid heat conducting medium is stated as

$$\frac{\mathrm{d}}{\mathrm{d}t}\int_V \varrho\varepsilon\,\mathrm{d}V = \int_V \varrho s\,\mathrm{d}V - \int_{\partial V} k\,\mathrm{d}S\,, \tag{26}$$

where t is time, ϱ is mass density, ε is internal energy per unit mass, V is volume, s is the heat source per unit mass und k is the heat flux per unit area of the boundary surface ∂V. By the standard tetrahedron argument, the existence of a heat flux vector q^i can be established such that

$$k = q^i n_i \tag{27}$$

where n_i are the components of the exterior unit normal to ∂V. Using Eq. (27) and the divergence theorem, Eq. (26) can be localized to yield

$$\varrho\dot{\varepsilon} = \varrho s - q^i_{,i}\,, \tag{28}$$

where a superimposed dot indicates the (material) time derivative $\frac{\mathrm{d}}{\mathrm{d}t}$, and where a comma before a subscript represents covariant differentiation in space.

Multiplying Eq. (28) by an arbitrary variation of the type

$$\delta\theta = \delta\theta_0 + z^I\,\delta\theta_I \tag{29}$$

and integrating yields the global equations

$$\frac{d}{dt}\int\limits_S \varrho_S e \, dS = \int\limits_S \varrho_S r \, dS - \int\limits_{\partial S} Q^\alpha n_\alpha \, dc \tag{30}$$

and

$$\frac{d}{dt}\int\limits_S \varrho_S e^I \, dS = \int\limits_S \varrho_S r^I \, dS + \int\limits_S Q^I \, dS - \int\limits_{\partial S} Q^{\alpha I} n_\alpha \, dc \, , \tag{31}$$

where the following notation has been introduced

$$\varrho_S = \int \varrho J \, dz \, , \qquad \varrho_S e = \int \varrho \varepsilon J \, dz \, , \qquad \varrho_S e^I = \int \varrho \varepsilon z^I J \, dz \, , \qquad (32), (33) (34)$$

$$\varrho_S r = \int \varrho s J \, dz - (q^3 J)|_i^o \, , \qquad \varrho_S r^I = \int \varrho s z^I J \, dz - (q^3 z^I J)|_i^o \, , \qquad (35), (36)$$

$$Q^\alpha = \int q^\alpha J \, dz \, , \qquad Q^{\alpha I} = \int q^\alpha z^I J \, dz \, , \qquad Q^I = \int q^3 J \, dz^I \qquad (37), (38), (39)$$

and the indices o and i indicate "outer" and "inner" surface limits, and where lower case Greek indices range from 1 to 2, and n_α are the components of the exterior unit normal to the boundary δS of the reference surface. The symbol J appearing in Eqs. (32) ... (39) denotes the known ratio between the elements of area away from and at the reference surface. Note that the radiation terms r and r^I incorporate the boundary values of the normal component of the heat flux vector at the outer and inner surfaces of the shell, much in the same way as the surface "body" force in a shell incorporates the surface tractions acting on those same surfaces.

Localizing Eqs. (30) and (31) by means of the surface divergence theorem and denoting by a vertical bar covariant differentiation with respect to the surface metric yields

$$\varrho_S \dot{e} = \varrho_S r - Q^\alpha|_\alpha \tag{40}$$

and

$$\varrho_S \dot{e}^I = \varrho_S r^I + Q^I - Q^{\alpha I}|_\alpha \tag{41}$$

which represent $N + 1$ equations of energy balance for the $N + 1$ temperature variables.

As before, a more compact formulation is obtained by using an additional coordinate z_0 and letting capital Greek indices range from 0 to N. Thus, Eqs. (29), (33)...(39) become

$$\delta\theta = \delta\theta_\Lambda z^\Lambda \, , \qquad \varrho_S e^\Lambda = \int \varrho \varepsilon z^\Lambda J \, dz \, , \qquad (42), (43)$$

$$\varrho_S r^\Lambda = \int \varrho s z^\Lambda J \, dz - (q^3 z^\Lambda J)|_i^o \, , \qquad Q^{\alpha\Lambda} = \int q^\alpha z^\Lambda J \, dz \, , \qquad Q^\Lambda = \int q^3 J \, dz^\Lambda \, , \qquad (44), (45), (46)$$

where Eq. (22) implies

$$Q^0 = 0 \tag{47}$$

since $dz^0 = 0$.

Finally, Eqs. (40) and (41) coalesce into the single equation

$$\varrho_S \dot{e}^\Lambda = \varrho_S r^\Lambda + Q^\Lambda - Q^{\alpha\Lambda}|_\alpha \, . \tag{48}$$

4. Entropy Production

Assume the validity of the classical Clausius-Duhem inequality in the form

$$\frac{d}{dt}\int\limits_V \varrho\eta \, dV \geqq \int\limits_V \varrho s\theta \, dV - \int\limits_{\partial V} k\theta \, dS \tag{49}$$

where η is the entropy per unit mass, and θ is the coldness (inverse of the temperature).

In order to approximate this three dimensional inequality we will require that it hold globally and independently for each layer. Thus for the I-th layer, assumed of constant thickness, we get

$$\frac{d}{dt}\int\limits_S\int\limits_{h_I} \varrho\eta J \, dz^I \, dS \geqq \int\limits_S\int\limits_{h_I} \varrho s\theta J \, dz^I \, dS - \int\limits_{\partial S}\int\limits_{h_I} q^\alpha n_\alpha \theta J \, dz^I \, dc - \frac{l}{h_I}\int\limits_S (q^3\theta J)|_{(I-1)\text{-th interface}}^{I\text{-th interface}} \, dS \, . \tag{50}$$

So as to be able to exploit the N inequalities, Eq. (50), it is convenient to express them, as well as the energy balance equations, in terms of a common set of two-dimensional variables. One could choose any of the sets discussed in Section 2, since they are all equally adequate for the characterization of piece-wise linear functions.

We choose the coefficients s_Λ, θ_Λ, q_Λ^i and ε_Λ of the expansion of s, θ, q^i and ε given by Eqs. (1) as our basic variables and introduce the following notation for two particular cases of Eq. (24) and (25), namely,

$$D^{\Lambda\Delta} \equiv S_{(J)}^{\Lambda\Delta} \, , \qquad D^{\Delta\Gamma\Lambda} \equiv C_{(J)}^{\Delta\Gamma\Lambda} \, ,$$

and

$$E^{\Lambda\Delta} \equiv S_{(\varrho J)}^{\Lambda\Delta} \, , \qquad E^{\Delta\Gamma\Lambda} \equiv C_{(\varrho J)}^{\Delta\Gamma\Lambda} \, . \tag{51}$$

The entropy inequality, Eq. (50), then becomes

$$\frac{\mathrm{d}}{\mathrm{d}t}\int\limits_S \eta^I \,\mathrm{d}S \geqq \int\limits_S E^{\Lambda\Delta I}\theta_\Lambda s_\Delta \,\mathrm{d}S - \int\limits_{\partial S} D^{\Lambda\Delta I} q^\alpha_\Lambda \theta_\Delta n_\alpha \,\mathrm{d}c - \frac{l}{h_I}\int\limits_S q^3_\Lambda \theta_\Delta [Z^\Lambda_I Z^\Delta_I J_I - Z^\Lambda_{I-1} Z^\Delta_{I-1} J_{I-1}] \,\mathrm{d}S \tag{52}$$

where

$$\eta^I = \int \varrho \eta J \,\mathrm{d}z^I , \tag{53}$$

and where J_I is the value of J at the I-th interface. The local version of Eq. (52) is

$$\dot{\eta}^I \geqq E^{\Lambda\Delta I}\theta_\Lambda s_\Delta - (D^{\Lambda\Delta I} q^\alpha_\Lambda \theta_\Delta)|_\alpha - \frac{l}{h_I} q^3_\Lambda \theta_\Delta (Z^\Lambda_I Z^\Delta_I J_I - Z^\Lambda_{I-1} Z^\Delta_{I-1} J_{I-1}) . \tag{54}$$

In order to get an approximate set of restrictions on constitutive equations we read off the radiation term s_Δ from the energy balance equations as

$$s_\Delta = \dot{\varepsilon}_\Delta + E^{-1}_{\Delta\Lambda}\{q^3_\Gamma (Z^\Lambda_N Z^\Gamma_N J_N - Z^\Lambda_0 Z^\Gamma_0 J_0) - q^3_\Gamma D^{0\Gamma\Lambda} + (q^\alpha_\Gamma D^{\Lambda\Gamma})|_\alpha\} , \tag{55}$$

and introduce the result into the entropy inequalities, to yield the N inequalities

$$\dot{\eta}^I \geqq E^{\Lambda\Delta I}\theta_\Lambda \dot{\varepsilon}_\Delta + E^{\Lambda\Delta I}\theta_\Lambda E^{-1}_{\Delta\Phi}\{q^3_\Gamma (Z^\Phi_N Z^\Gamma_N J_N - Z^\Phi_0 Z^\Gamma_0 J_0) - q^3_\Gamma D^{0\Gamma\Phi} + (q^\alpha_\Gamma D^{\Phi\Gamma})|_\alpha\} -$$
$$- (D^{\Lambda\Delta I} q^\alpha_\Lambda \theta_\Delta)|_\alpha - \frac{l}{h_I} q^3_\Lambda \theta_\Lambda (Z^\Lambda_I Z^\Delta_I J_I - Z^\Lambda_{I-1} Z^\Delta_{I-1} J_{I-1}) . \tag{56}$$

Thus assuming

$$\eta^I = \eta^I(\theta_\Lambda, \theta_{\Lambda,\alpha}) , \qquad \varepsilon_\Delta = \varepsilon_\Delta(\theta_\Lambda, \theta_{\Lambda,\alpha}) , \tag{57}$$

we get the $2N$ restrictions

$$\frac{\partial \eta^I}{\partial \theta_\Gamma} = E^{\Lambda\Delta I}\theta_\Lambda \frac{\partial \varepsilon_\Delta}{\partial \theta_\Lambda}, \qquad \frac{\partial \eta^I}{\partial \theta_{\Gamma,\alpha}} = E^{\Lambda\Delta I}\theta_\Lambda \frac{\partial \varepsilon_\Delta}{\partial \theta_{\Lambda,\alpha}} \tag{58), (59}$$

and the N residual inequalities

$$0 \geqq E^{\Lambda\Delta I}\theta_\Lambda E^{-1}_{\Delta\Phi}\{q^3_\Gamma (Z^\Phi_N Z^\Gamma_N J_N - Z^\Phi_0 Z^\Gamma_0 J_0) - q^3_\Gamma D^{0\Gamma\Phi} + (q^\alpha_\Gamma D^{\Phi\Gamma})|_\alpha\} - \frac{l}{h_I} q^3_\Lambda \theta_\Delta (Z^\Lambda_I Z^\Delta_I J_I - Z^\Lambda_{I-1} Z^\Delta_{I-1} J_{I-1}) . \tag{60}$$

It is interesting to note that introducing a layer free energy variable as

$$\psi^I = E^{\Lambda\Delta I}\varepsilon_\Lambda \theta_\Delta - \eta^I . \tag{61}$$

Eq. (59) simply states that

$$\frac{\partial \psi^I}{\partial \theta_{\Gamma,\alpha}} = 0 \tag{62}$$

and Eq. (58) becomes

$$\frac{\partial \psi^I}{\partial \theta_\Gamma} = E^{\Lambda\Gamma I}\varepsilon_\Lambda . \tag{63}$$

For the case of one layer with the middle surface as reference, Eqs. (58), (59) and (60) reduce essentially to GROT's formulation [2].

5. Conclusions

A theory of rigid non-isothermal shells is derived by dividing the shell into a number of layers and approximating the relevant fields by means of piece-wise linear functions through the shell thickness. The exact entropy inequality is converted into a number of inequalities, one for each layer, in terms of layer entropy densities, and thus restrictions on constitutive equations are obtained. Although an alternative approach, used by GREEN and NAGHDI [4], based on taking moments of all orders of the three-dimensional equations, seems not to be general enough to enforce the entropy inequality, a recent paper by the same authors [9] may provide a satisfactory solution through the introduction of an equation of entropy balance.

The present formulation reduces, in the case of a single layer, to a particular case of GROT's [2] theory of media with microstructure. On the other hand, it is not obvious how to apply the general method and results of this work to a similar theory of general oriented media, a problem which remains a challenge to researchers.

References

1 TIMOSHENKO, S. and GOODIER, J. N., Theory of Elasticity, 2nd Ed., McGraw-Hill, 1957.
2 GROT, R. A., Thermodynamics of a Continuum with Microstructure, Int. J. Engng. Sci., 7, 801–814 (1969).
3 KOH, S. L. and WENG, M.-T., Thermoelasticity of Materials with Microstructure, Letters Appl. Engng. Sci., 4, 29–41 (1976).

4 Green, A. E. and Naghdi, P. M., Non-Isothermal Theory of Rods, Plates and Shells, Int. J. Solids Structures, **6**, 209–244 (1970).
5 Krätzig, W. B., Allgemeine Schalentheorie beliebiger Werkstoffe und Verformungen, Ingenieur-Archiv, **40**, 311–326 (1971).
6 Biot, M. A., Variational Principles in Heat Transfer, Oxford University Press, London, 1970.
7 Epstein, M. and Glockner, P. G., Nonlinear Analysis of Multilayered Shells, Int. J. Solids Structures, **13**, 1081–1089 (1977).
8 Epstein, M. and Murray, D. W., Three-Dimensional Large Deformation Analysis of Thin-Walled Beams, Int. J. Solids and Structures, **12**, 867–876 (1976).
9 Green, A. E. and Naghdi, P. M., On Thermodynamics and the Nature of the Second Law, Proc. Roy. Soc., London, **357** A, 253–270 (1977).

Eingereicht am 31. 1. 1979, in revidierter Fassung am 25. 5. 1979

Anschrift: Prof. P. G. Glockner, Department of Mechanical Engineering, The University of Calgary, Calgary, Alberta, Canada T2N 1N 4

ZAMM 60, 71–82 (1980)

J. J. TELEGA

Variational Principles for Rate Boundary-Value Problems in Non-Associated Plasticity

This study concerns a systematic search for variational principles applicable in solving rate boundary-value problems of nonassociated plasticity. An investigated solid is assumed to be elastic-plastic, work hardening and/or softening, obeying non-associated flow rule. Coupling between elastic and plastic deformation has also been included. It is shown by means of the method of adding the adjoint operator that the counterparts of the variational principles of Hu-Washizu and Hellinger-Reissner can be derived. Variational principles corresponding to the principles of a stationary value of potential energy and stationary value of complementary energy are studied. Convenient in applications counterparts of the principles of virtual velocities, virtual changes of rate of stresses and virtual work are derived. Two reciprocal theorems are formulated.

Die Arbeit ist der systematischen Suche nach Variationsprinzipien gewidmet, die bei der Lösung von Randwertproblemen auf dem Gebiet der nicht-assoziierten Plastizität nutzbar sind. Der untersuchte Körper wird als elastisch-plastisch, verfestigend und/oder erweichend angenommen, wobei die Fließbedingung mit dem Stoffgesetz nicht assoziiert. Eine Kopplungzwischen elastischen und plastischen Deformationen wurde ebenfalls berücksichtigt. Mittels des adjungierten Operators wird gezeigt, daß die Gegenstücke der Variationsprinzipien von Hu-Washizu und Hellinger-Reissner hergeleitet werden können. Es werden Variationsprinzipien studiert, die den Prinzipien eines stationären Wertes der potentiellen Energie und eines stationären Wertes der komplementären Energie entsprechen. Im Hinblick auf Anwendungen wurden die Gegenstücke der Prinzipien der virtuellen Geschwindigkeiten, der virtuellen Spannungen und der virtuellen Arbeit hergeleitet. Zwei reziproke Theoreme werden formuliert.

Работа посвящена систематическому поиску вариационных принципов, применяемых при решении задач с граничными условиями в области неассоцированной пластичности. Исследованное твёрдое тело предполагается упруго-пластичным, с упрочнением и/или с пластификацией, подчиняющимся неассоцированному условию текучести. Учитывается также связь между упругими и пластическими деформациями. Посредством сопряжённого оператора показывается, что могут быть выведены принципы, подобные вариационным принципам Ху-Вашицу и Хеллингера-Рейснера. Изучаются вариационные принципы, соответствующие принципам стационарного значения потенциальной энергии и стационарного значения дополнительной энергии. В расчёте на приложения выводятся принципы, подобные принципам виртуальных скоростей, виртуальных напряжений и виртуальной работы. Формулируются две обратные теоремы.

1. Introduction

This paper is concerned with a systematic search for variational principles applicable to solving rate boundary-value problems of non-associated plasticity. We shall investigate elastic-plastic, work hardening and/or softening solids, described by non-associated flow laws [1–5]. Coupling between elastic and plastic deformations has also been included [6–9].

It seems that non-associated flow laws are necessary to describe properly some phenomena encountered in plastic behavior of solids and especially of soils and rocks. For instance, experiments reveal deviation of the plastic strain increment vector from the outward normal to a yield surface in the case of plastic flow of soils [10, 11]. Though theories of materials obeying non-associated flow laws are known for years (cf. [1–12] where further references are cited), nevertheless mathematical methods are not so well developped as in the case of associated plasticity. We think this is due mainly to the fact that a boundary-value operator is not self-adjoint with respect to the Cartesian bilinear form (see Section 3 of this paper).

Among mathematical methods variational principles are very important and useful. Therefore we shall derive several variational principles regarding solids described by non-associated flow laws.

Making use of the method of adding the adjoint operator [13], we formulate four variational principles. An interpretation of the adjoint system is delivered. An alternative approach for obtaining variational principles is proposed. This approach is based on the principle of virtual velocities and on the principle of virtual changes of rate of stresses for "an original and adjoint solid treated as a whole".

An attempt is made to extend the reciprocal theorem, formulated in [14] for associated flow law, to non-associated flow rules.

2. Equations of the considered elastic-plastic solids obeying non-associated flow laws

2.1. We recall in this section some results derived in [1, 2].

Let $\dot{\boldsymbol{\varepsilon}} = (\dot{\varepsilon}_{ij})$, $\dot{\boldsymbol{\sigma}} = (\dot{\sigma}_{ij})$, $\boldsymbol{\sigma} = (\sigma_{ij})$ denote the rate of deformation, rate of stress and stress tensors, respectively. In this paper the range of latin indices is 1, 2, 3, unless otherwise stated. Let $f = f(\sigma_{ij}, \beta_{ij}^{(s)}, q^{(r)}) = 0$ be a yield condition. Tensors $\boldsymbol{\beta}^{(s)}$, $s = 1, \dots, N_1$, and scalars $q^{(r)}$, $r = 1, \dots, N_2$ characterize changes of the yield surface during deformation. We assume that at every point of a yield surface a unit, normal vector $\boldsymbol{n}_f = (n_{ij})$ exists.

The linear, constitutive rate equations, considered in [1, 2] have the form

$$\begin{aligned} \dot\varepsilon_{ij} &= M_{ijkl}\dot\sigma_{kl} + \frac{1}{H} g_{ij} f_{kl}\dot\sigma_{kl}, \quad \text{for } f_{kl}\dot\sigma_{kl} > 0, \ f = 0, \\ \dot\varepsilon_{ij} &= M_{ijkl}\dot\sigma_{kl}, \quad \text{for } f_{kl}\dot\sigma_{kl} \leqq 0, \ f = 0; \ \text{or } f < 0, \end{aligned} \tag{2.1}$$

where $\boldsymbol{M} = (M_{ijkl})$ is the elasticity tensor, $f_{ij} = \partial f/\partial \sigma_{ij}$, H denotes the work-hardening parameter, $g_{ij} = g_{ij}(\sigma_{ij}, \beta_{ij}^{(s)}, q^{(r)})$. For further considerations we assume that a plastic potential $g = g(\sigma_{ij}, \beta_{ij}^{(s)}, q^{(r)})$ exists, that is $g_{ij} = \partial g/\partial \sigma_{ij}$. In this case the relation (2.1) is identical with constitutive equations proposed by MELAN [3].

If we put

$$A_{ijkl} = \begin{cases} M_{ijkl} + \dfrac{1}{H} g_{ij} f_{kl}, & \text{for } f_{kl}\dot\sigma_{kl} > 0, \ f = 0 \\ M_{ijkl}, & \text{for } f_{kl}\dot\sigma_{kl} \leqq 0, \ f \doteq 0; \ \text{or } f < 0 \end{cases} \tag{2.2}$$

then (2.1) takes the form

$$\dot\varepsilon_{ij} = A_{ijkl}\dot\sigma_{kl}. \tag{2.3}$$

We see that, for $g \neq f$, A_{ijkl} is not symmetric with respect to the group of indices ij and kl.

The inversion of the relation (2.3) leads to

$$\dot\sigma_{ij} = B_{ijkl}\dot\varepsilon_{kl}, \tag{2.4}$$

where

$$B_{ijkl} = \begin{cases} L_{ijkl} - \dfrac{L_{ijpq} m_{pq} L_{rskl} n_{rs}}{K + L_{abcd} n_{ab} m_{cd}}, & \text{for } f = 0, \ L_{ijkl}\dot\varepsilon_{kl} n_{ij} \geqq 0 \\ L_{ijkl}, & \text{for } L_{ijkl}\dot\varepsilon_{kl} n_{ij} < 0, \ f = 0; \ \text{or } f < 0. \end{cases} \tag{2.5}$$

Here

$$\boldsymbol{L} = \boldsymbol{M}^{-1}, \qquad K = \frac{H}{(g_{ij} g_{ij})^{1/2} (f_{kl} f_{kl})^{1/2}},$$

$$m_{ij} = \frac{g_{ij}}{(g_{kl} g_{kl})^{1/2}}, \qquad n_{ij} = \frac{f_{ij}}{(f_{kl} f_{kl})^{1/2}};$$

$\boldsymbol{n}_g = (m_{ij})$ is an outward unit vector, normal to a surface representing the plastic potential in the nine-dimensional Euclidean space.

We proceed to proving that the constitutive equations considered in [5, 7—9] are essentially of the form (2.3), (2.4).

2.2. Let $f(\boldsymbol{\sigma}, \boldsymbol{\alpha})$ be a yield function and $g(\boldsymbol{\sigma}, \boldsymbol{\alpha})$ a plastic potential, where $\boldsymbol{\alpha} = (\alpha_\varrho)$, $\varrho = 1, \ldots, N$ represents internal variables [5].

Denote by Ω_p the plasticized part of a considered body $\Omega \subset \mathbb{R}^3$, where $f(\boldsymbol{\sigma}, \boldsymbol{\alpha}) = 0$. We have [5]

$$\dot{\boldsymbol{\sigma}} = \boldsymbol{L}\dot{\boldsymbol{\varepsilon}} - \lambda \boldsymbol{L} \frac{\partial g}{\partial \boldsymbol{\sigma}}, \qquad \lambda \geqq 0 \tag{2.6}$$

$$\dot{\boldsymbol{\alpha}} = \lambda \boldsymbol{l}(\boldsymbol{\sigma}, \boldsymbol{\alpha}), \tag{2.7}$$

Here

$$\lambda = \begin{cases} \dfrac{\dfrac{\partial f}{\partial \boldsymbol{\sigma}} \boldsymbol{L}\dot{\boldsymbol{\varepsilon}}}{\dfrac{\partial f}{\partial \boldsymbol{\sigma}} \boldsymbol{L} \dfrac{\partial g}{\partial \boldsymbol{\sigma}} - \dfrac{\partial f}{\partial \boldsymbol{\alpha}} \boldsymbol{l}}, & \text{if } f = 0, \ \dot f \doteq 0, \\ 0, \ \text{if } f < 0; \ \text{or } f = 0, \ \dot f < 0, \end{cases} \tag{2.8}$$

and $\boldsymbol{L} = (L_{ijkl})$; $i, j, k, l = 1, 2, 3$ is the tensor of elastic moduli, whereas $\dot f = (\partial f/\partial \boldsymbol{\sigma})\, \dot{\boldsymbol{\sigma}} + (\partial f/\partial \boldsymbol{\alpha})\, \dot{\boldsymbol{\alpha}}$.

To avoid confusion, both absolute and indices notations are used. Taking account of (2.8) in (2.6), (2.7) we arrive at

$$\dot{\boldsymbol{\sigma}} = \begin{cases} \left(\boldsymbol{L} - \dfrac{\left(\boldsymbol{L}\dfrac{\partial g}{\partial \boldsymbol{\sigma}}\right)\left(\dfrac{\partial f}{\partial \boldsymbol{\sigma}}\boldsymbol{L}\right)}{\dfrac{\partial f}{\partial \boldsymbol{\sigma}} \boldsymbol{L} \dfrac{\partial g}{\partial \boldsymbol{\sigma}} - \dfrac{\partial f}{\partial \boldsymbol{\alpha}} \boldsymbol{l}} \right) \dot{\boldsymbol{\varepsilon}} & \text{if } f = 0, \ \dot f = 0, \\ \boldsymbol{L}\dot{\boldsymbol{\varepsilon}}, \ \text{if } f < 0; \ \text{or } f = 0, \ \dot f < 0 \end{cases} \tag{2.9}$$

$$\dot{\boldsymbol{\alpha}} = \begin{cases} \dfrac{\boldsymbol{l} \dfrac{\partial f}{\partial \boldsymbol{\sigma}} \boldsymbol{L}\dot{\boldsymbol{\varepsilon}}}{\dfrac{\partial f}{\partial \boldsymbol{\sigma}} \boldsymbol{L} \dfrac{\partial g}{\partial \boldsymbol{\sigma}} - \dfrac{\partial f}{\partial \boldsymbol{\alpha}} \boldsymbol{l}} & \text{if } f = 0, \ \dot f = 0, \\ 0, \ \text{if } f < 0; \ \text{or } f = 0, \ \dot f < 0. \end{cases} \tag{2.10}$$

Thus we can write

$$\dot{\sigma}_{ij} = B_{ijkl}\dot{\varepsilon}_{kl}\,, \tag{2.11}$$

$$\dot{\alpha}_\varrho = l_\varrho C_{kl}\dot{\varepsilon}_{kl}\,, \tag{2.12}$$

where

$$B_{ijkl} = \begin{cases} L_{ijkl} - \dfrac{L_{ijab}\dfrac{\partial g}{\partial \sigma_{ab}}\dfrac{\partial f}{\partial \sigma_{rs}}L_{rskl}}{\dfrac{\partial f}{\partial \sigma_{mn}}L_{mnpq}\dfrac{\partial g}{\partial \sigma_{pq}} - \dfrac{\partial f}{\partial \alpha_\varrho}l_\varrho} & \text{if } f = 0\,,\ \dot{f} = 0\,, \\ L_{ijkl}\,, \ \text{if } f < 0; \ \text{or } f = 0\,,\ \dot{f} < 0 \end{cases} \tag{2.13}$$

$$C_{kl} = \begin{cases} \dfrac{\dfrac{\partial f}{\partial \sigma_{ij}}L_{ijkl}}{\dfrac{\partial f}{\partial \sigma_{mn}}L_{mnpq}\dfrac{\partial g}{\partial \sigma_{pq}} - \dfrac{\partial f}{\partial \alpha_\varrho}l_\varrho} & \text{if } f = 0\,,\ \dot{f} = 0\,, \\ 0\,, \ \text{if } f < 0; \ \text{or } f = 0\,,\ \dot{f} < 0\,. \end{cases} \tag{2.14}$$

The matrix $\boldsymbol{B} = (B_{ijkl})$ is piecewise constant throughout the plastic domain and reduces to $\boldsymbol{L}$ in the elastic domain. To arrive at variational formulations it is thus appropriate to assume $B_{ijkl} \in L^\infty(\Omega)$ (for a definition of L^∞ — spaces, cf. [15]).

From (2.12), we conclude that the rate $\boldsymbol{\alpha}$ of internal variables can be calculated once the rates of deformations are known, $\boldsymbol{\sigma}$ and $\boldsymbol{\alpha}$ being prescribed. Thus in variational formulations $\dot{\boldsymbol{\alpha}}$ has not to be taken into account, what significantly simplifies such formulations.

The function

$$H = -\frac{\partial f}{\partial \boldsymbol{\alpha}}\boldsymbol{l} \tag{2.15}$$

is a work hardening, or softening, modulus. For a more elaborate analysis, cf. for instance [2, 4, 8].

It is possible to invert the relation (2.11). From $\dot{\boldsymbol{\varepsilon}}^p = \lambda(\partial g/\partial \boldsymbol{\sigma})$, $\dot{f} = 0$ and (2.7) we obtain

$$\lambda = -\frac{1}{\dfrac{\partial f}{\partial \boldsymbol{\alpha}}\boldsymbol{l}}\left(\frac{\partial f}{\partial \boldsymbol{\sigma}}\dot{\boldsymbol{\sigma}}\right). \tag{2.16}$$

Taking account of (2.16) in $\dot{\boldsymbol{\varepsilon}} = \boldsymbol{M}\dot{\boldsymbol{\sigma}} + \lambda(\partial g/\partial \boldsymbol{\sigma})$ we eventually arrive at

$$\dot{\varepsilon}_{ij} = A_{ijhl}\dot{\sigma}_{hl}$$

where

$$A_{ijhl} = \begin{cases} M_{ijhl} - \dfrac{1}{\dfrac{\partial f}{\partial \alpha_\varrho}l_\varrho}\dfrac{\partial f}{\partial \sigma_{hl}}\dfrac{\partial g}{\partial \sigma_{ij}}\,, & \text{if } f = 0\,,\ \dot{f} = 0 \\ M_{ijhl}\,, \ \text{if } f < 0; \ \text{or } f = 0\,,\ \dot{f} < 0\,. \end{cases} \tag{2.17}$$

2.3. Behaviour of materials exhibiting coupling of elastic and plastic deformations has been considered in [7—9]. For our purposes we shall slightly modify the equations given in the above mentioned papers. Similarly as in [5] we introduce internal variables. We have [7—9]

$$\dot{\boldsymbol{\varepsilon}} = \dot{\boldsymbol{\varepsilon}}^e + \dot{\boldsymbol{\varepsilon}}^p = \dot{\boldsymbol{\varepsilon}}' + \dot{\boldsymbol{\varepsilon}}'' + \dot{\boldsymbol{\varepsilon}}^p \tag{2.18}$$

where

$$\dot{\boldsymbol{\varepsilon}}^p = \lambda\frac{\partial g}{\partial \boldsymbol{\sigma}}. \tag{2.19}$$

The reversible part $\dot{\boldsymbol{\varepsilon}}'$ of the elastic strain rate tensor expresses as follows

$$\dot{\varepsilon}'_{ij} = M_{ijhk}\dot{\sigma}_{hk} \tag{2.20}$$

and $M_{ijhk} = M_{ijhk}(\boldsymbol{\alpha})$. In [8, 9] the elastic compliance tensor $\boldsymbol{M} = (M_{ijkl})$ is assumed to be positive definite. The coupled term $\dot{\boldsymbol{\varepsilon}}''$ is the irreversible part of the elastic strain rates and reads

$$\dot{\varepsilon}''_{ij} = \lambda\frac{\partial M_{ijhk}}{\partial \alpha_\varrho}\sigma_{hk}l_\varrho\,. \tag{2.21}$$

Particularly, when the only internal variable is $\boldsymbol{\varepsilon}^p$, then

$$\dot{\varepsilon}''_{ij} = \lambda \frac{\partial M_{ijhk}}{\partial \varepsilon^p_{rs}} \sigma_{hk} \frac{\partial g}{\partial \sigma_{rs}} \tag{2.22}$$

and we recover the theory of HUECKEL and MAIER [8, 9].

Denote by Ω_p the plastic volume, where $f(\boldsymbol{\sigma}, \boldsymbol{\alpha}) = 0$. After NGUYEN and BUI [5] we assume the following definition: "A material point belonging to Ω_p is at the state of plastic loading if $\dot{f}(\boldsymbol{\sigma}, \boldsymbol{\alpha}) = 0$, and at the state of elastic unloading if $\dot{f}(\boldsymbol{\sigma}, \boldsymbol{\alpha}) < 0$".

From (2.20) and the assumption of positive definiteness of $\boldsymbol{M}$ we obtain

$$\dot{\sigma}_{ij} = L_{ijkl}\dot{\varepsilon}'_{kl} \tag{2.23}$$

where $\boldsymbol{L}$ is the inverse of $\boldsymbol{M}$ and $L_{ijhk} = L_{ijhk}(\boldsymbol{\alpha}_\varrho)$, (cf. [6--7]). From (2.18), (2.19), (2.21) and (2.23) we finally get

$$\lambda = \frac{\dfrac{\partial f}{\partial \sigma_{ij}} L_{ijkl}\dot{\varepsilon}_{kl}}{\dfrac{\partial f}{\partial \sigma_{ab}} L_{abcd} \dfrac{\partial M_{cdmn}}{\partial \alpha_\varrho} \sigma_{mn} l_\varrho + \dfrac{\partial f}{\partial \sigma_{mn}} L_{mnrs} \dfrac{\partial g}{\partial \sigma_{rs}} - \dfrac{\partial f}{\partial \alpha_\varrho} l_\varrho}. \tag{2.24}$$

Assuming that the denominator in (2.24) is positive then the condition of plastic loading $\dot{f}(\boldsymbol{\sigma}, \boldsymbol{\alpha}) = 0$ is equivalent to $\dfrac{\partial f}{\partial \boldsymbol{\sigma}} \boldsymbol{L}\dot{\boldsymbol{\varepsilon}} \geqq 0$, (cf. [5]).

The constitutive equation has eventually the form

$$\dot{\sigma}_{ij} = B_{ijkl}\dot{\varepsilon}'_{kl} \tag{2.25}$$

and now

$$B_{ijkl} = \begin{cases} L_{ijkl} - \dfrac{L_{ijmn}\left(\dfrac{\partial M_{mnab}}{\partial \alpha_\varrho} \sigma_{ab} l_\varrho + \dfrac{\partial g}{\partial \sigma_{mn}}\right) \dfrac{\partial f}{\partial \sigma_{pq}} L_{pqkl}}{\dfrac{\partial f}{\partial \sigma_{ab}} L_{abcd} \dfrac{\partial M_{cdmr}}{\partial \alpha_\varrho} \sigma_{mn} l_\varrho + \dfrac{\partial f}{\partial \sigma_{mn}} L_{mnrs} \dfrac{\partial g}{\partial \sigma_{rs}} - \dfrac{\partial f}{\partial \alpha_\varrho} l_\varrho}, & \text{if } f = 0, \ \dot{f} = 0, \\ L_{ijkl}, & \text{if } f < 0; \text{ or } f = 0, \ \dot{f} < 0. \end{cases} \tag{2.26}$$

Both in (2.11) and (2.25) we have in general $B_{ijkl} \neq B_{klij}$.

We note that the sum $\dot{\boldsymbol{\varepsilon}}^p + \dot{\boldsymbol{\varepsilon}}''$ of irreversible strain rates is directed in general neither along the outward normal to the yield surface, nor along the gradient of plastic potential.

Let us invert the relation (2.25). From (2.18)—(2.21) and (2.16) we finally obtain

$$\dot{\varepsilon}_{ij} = A_{ijkl}\dot{\sigma}_{kl} \tag{2.27}$$

where

$$A_{ijkl} = \begin{cases} M_{ijkl} - \dfrac{1}{\dfrac{\partial f}{\partial \alpha_\varrho} l_\varrho} \left(\dfrac{\partial M_{ijmn}}{\partial \alpha_\varrho} l_\varrho \sigma_{mn} + \dfrac{\partial g}{\partial \sigma_{ij}}\right) \dfrac{\partial f}{\partial \sigma_{kl}}, & \text{if } f = 0, \ \dot{f} = 0, \\ M_{ijkl}, \ \text{if } f < 0; \text{ or } f = 0, \ \dot{f} < 0. \end{cases} \tag{2.28}$$

Remark 2.1. The representations (2.11) and (2.25) do not hold for instantaneous perfectly-plastic behaviour, because the plastic multiplier λ is then not expressed by (2.8), or (2.24).

3. Derivation of the form of the adjoint operator

Let Ω be a bounded domain occupied by the considered elastic-plastic solid, in the three-dimensional Euclidean space; $S = \partial\Omega$, $S = S_1 \cup S_2$. On the parts S_1, S_2 of the boundary velocity $\bar{\boldsymbol{v}} = (v_i)$ and rate of traction $\boldsymbol{p} = (p_i)$ are prescribed, respectively. We assume that all functions appearing in the paper are sufficiently regular. The considerations which follow are *within the theory of small displacements.*

A complete system of equations describing rate boundary-value problem for elastic-plastic solid, obeying the non-associated flow law (2.4), or (2.11), or (2.25) can be written as follows

$$L(\dot{\boldsymbol{\sigma}}, \boldsymbol{v}, \dot{\boldsymbol{\varepsilon}}) = \boldsymbol{G} \Leftrightarrow \begin{cases} v_{(j,i)} - \dot{\varepsilon}_{ij} = 0, \ \Omega & (3.1) \\ -n_{(i}v_{j)} = -n_{(i}\bar{v}_{j)}, \ S_1 & (3.2) \\ -\dot{\sigma}_{ij,i} = b_j, \ \Omega & (3.3) \\ n_i\dot{\sigma}_{ij} = p_j, \ S_2 & (3.4) \\ -\dot{\sigma}_{ij} + B_{ijkl}\dot{\varepsilon}_{kl} = 0, & (3.5) \end{cases}$$

where L is a linear operator, $\boldsymbol{G}$ denotes the right-hand side of relations (3.1)—(3.5), (b_i) is the rate of body forces. The boundary condition (3.2) is equivalent to $v_i = \bar{v}_i$. Such form of the equations describing elastic-plastic flow is convenient in formulating variational principles (cf. [24, 25]).

Let U_1, U_2 be suitably chosen linear spaces, such that $(\dot{\boldsymbol{\sigma}}, \boldsymbol{v}, \dot{\boldsymbol{\varepsilon}}) = \boldsymbol{u} \in U_1$, $\boldsymbol{G} \in U_2$. Thus we have $L\colon U_1 \to U_2$. A domain of the operator L may be a properly chosen subspace of U_1.

Let on the Cartesian product $U_2 \times U_1$ a non-degenerate bilinear form $\langle .,. \rangle$ be prescribed (cf. [13, 16]). We denote by L^* the adjoint of the operator L, that is

$$\langle L\boldsymbol{u}_1, \boldsymbol{u}_2 \rangle = \langle L^*\boldsymbol{u}_2, \boldsymbol{u}_1 \rangle ; \qquad \boldsymbol{u}_1, \boldsymbol{u}_2 \in U_1 . \tag{3.6}$$

We say that the given linear operator L is self-adjoint, i.e. potential, with respect to the chosen bilinear form $\langle . , . \rangle$, iff $L = L^*$. For a self-adjoint operator a variational formulation can be given, that is a functional F exists such that $\delta F = \langle L\boldsymbol{u} - \boldsymbol{G}, \delta \boldsymbol{u} \rangle = 0 \Rightarrow L\boldsymbol{u} = \boldsymbol{G}$. Here the symbol "$\delta$" denotes the variation of a respective quantity.

We shall prove that the operator L, given by the relations (3.1)—(3.5), is not self-adjoint. It is sufficient to find out the adjoint operator L^*. Let $\boldsymbol{u}_\alpha = (\dot{\boldsymbol{\sigma}}^{(\alpha)}, \boldsymbol{v}^{(\alpha)}, \dot{\boldsymbol{\varepsilon}}^{(\alpha)})$, $\alpha = 1, 2$ be two arbitrary elements belonging to the domain of the operator L. As the form $\langle . , . \rangle$ we take the Cartesian bilinear form. Thus we obtain

$$\begin{aligned}\langle L\boldsymbol{u}_1, \boldsymbol{u}_2 \rangle &= \int_\Omega [(v_{j,i}^{(1)} - \dot{\varepsilon}_{ij}^{(1)})\, \dot{\sigma}_{ij}^{(2)} - \dot{\sigma}_{ij,i}^{(1)} v_j^{(2)} + (-\dot{\sigma}_{ij}^{(1)} + B_{ijkl}\dot{\varepsilon}_{kl}^{(1)})\, \dot{\varepsilon}_{ij}^{(2)}]\, \mathrm{d}\Omega - \\ &\quad - \int_{S_1} n_i v_j^{(1)} \dot{\sigma}_{ij}^{(2)}\, \mathrm{d}S + \int_{S_2} n_i \dot{\sigma}_{ij}^{(1)} v_j^{(2)}\, \mathrm{d}S .\end{aligned} \tag{3.7}$$

Integrating by parts and performing simple calculations, we get

$$\begin{aligned}\langle L\boldsymbol{u}_1, \boldsymbol{u}_2 \rangle &= \int_\Omega [(v_{j,i}^{(2)} - \dot{\varepsilon}_{ij}^{(2)})\, \dot{\sigma}_{ij}^{(1)} - \dot{\sigma}_{ij,i}^{(2)} v_j^{(1)} + (-\dot{\sigma}_{ij}^{(2)} + B_{klij}\dot{\varepsilon}_{kl}^{(2)})\, \dot{\varepsilon}_{ij}^{(1)}]\, \mathrm{d}\Omega - \\ &\quad - \int_{S_1} n_i v_j^{(2)} \dot{\sigma}_{ij}^{(1)}\, \mathrm{d}S + \int_{S_2} n_i \dot{\sigma}_{ij}^{(2)} v_j^{(1)}\, \mathrm{d}S = \langle L^*\boldsymbol{u}_2, \boldsymbol{u}_1 \rangle .\end{aligned} \tag{3.8}$$

The relation (3.8) results in the following form of the adjoint operator L^*

$$L^*(\dot{\boldsymbol{T}}, \boldsymbol{V}, \boldsymbol{E}) = \boldsymbol{G}^* \Leftrightarrow \begin{cases} V_{(j,i)} - \dot{E}_{ij} = 0,\ \Omega & (3.9)\\ -n_{(i}V_{j)} = -n_{(i}\bar{V}_{j)},\ S_1 & (3.10)\\ -\dot{T}_{ij,i} = B_j,\ \Omega & (3.11)\\ n_i\dot{T}_{ij} = P_j,\ S_2 & (3.12)\\ -\dot{T}_{ij} + B_{klij}\dot{E}_{kl} = 0 . & (3.13)\end{cases}$$

Arguments of the operator L^* are denoted by capital letters. We conclude that for the non-associated flow laws the operator L is not self-adjoint, because $B_{ijkl} \neq B_{klij}$. Equality $L = L^*$ holds only iff $B_{ijkl} = B_{klij}$.

The relations (3.9)—(3.13) describe the "adjoint solid" subjected to the boundary conditions (3.10), (3.12) and rate of body forces $\boldsymbol{B} = (B_i)$. The relation (3.13) may be called the constitutive equation for the adjoint solid. If the original solid is the one considered in Section 2.2, then for the adjoint solid the yield function f plays the role of the plastic potential. In this case we arrive finally at

$$\dot{T}_{kl} = L_{ijkl}\dot{E}_{ij} - \lambda^* L_{ijkl} \frac{\partial f}{\partial \sigma_{ij}} \tag{3.14}$$

where

$$\lambda^* = \begin{cases} \dfrac{\dfrac{\partial g}{\partial \boldsymbol{\sigma}} \boldsymbol{L}\dot{\boldsymbol{E}}}{\dfrac{\partial f}{\partial \boldsymbol{\sigma}} \boldsymbol{L} \dfrac{\partial g}{\partial \boldsymbol{\sigma}} - \dfrac{\partial f}{\partial \boldsymbol{\alpha}} \boldsymbol{l}}, & \text{if } f = 0,\ \dot{f} = 0 \\ 0, \ \text{if } f < 0; & \text{or } f = 0,\ \dot{f} < 0 . \end{cases} \tag{3.15}$$

Thus we have

$$\boldsymbol{E}^P = \lambda^* \frac{\partial f}{\partial \boldsymbol{\sigma}}, \tag{3.16}$$

E^P being the plastic part of $\dot{\boldsymbol{E}}$. It is reasonable to assume $\lambda^* \geqq 0$, this condition being equivalent to $(\partial g/\partial \boldsymbol{\sigma})\, \boldsymbol{L}\dot{\boldsymbol{E}} \geqq 0$. Further we conclude that the loading and unloading conditions are analogous both for the original and adjoint solids.

In the case of an original solid considered in Section 2.1. we must just replace $-\dfrac{\partial f}{\partial \boldsymbol{\alpha}} \boldsymbol{l}$ by H.

The formula (3.15) for λ^* at a first look may seem strange, but we should bear in mind that the adjoint solid plays an auxiliary role, rendering the whole problem self-adjoint, c.f. the next Section.

In the case of coupling between elastic and plastic deformation (Section 2.3), we obtain

$$\dot{E}^p_{ij} = \frac{\dfrac{\partial g}{\partial \sigma_{mn}} L_{klmn} \dot{E}_{kl}}{\Lambda} \frac{\partial f}{\partial \sigma_{ij}}, \tag{3.17}$$

$$\dot{E}''_{ij} = \frac{1}{\Lambda} \frac{\partial M_{mnab}}{\partial \alpha_\varrho} \sigma_{ab} l_\varrho L_{cdmn} \dot{E}_{cd} \frac{\partial f}{\partial \sigma_{ij}}, \tag{3.18}$$

where Λ denotes the denominator of the expression (2.24). We conclude that for the adjoint solid the irreversible part $\dot{\boldsymbol{E}}^p + \dot{\boldsymbol{E}}''$ is directed along the yield locus gradient.

Remark 3.1. Return to the situation of Section 2.1. Let Π_f, Π_g denote hyperplanes normal to $\boldsymbol{n}_f$ and $\boldsymbol{n}_g$, respectively (Fig. 1).

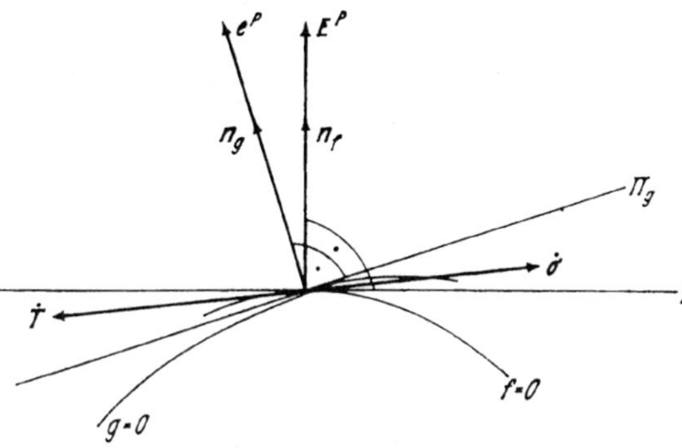

Fig. 1

We know that $\dot{\boldsymbol{E}}^p \perp \Pi_f$, $\dot{\boldsymbol{\varepsilon}}^p \perp \Pi_g$, where the superscript "p" indicates the plastic part of rate of deformations tensors. The following inequalities follow

$$\dot{\sigma}_{ij} \dot{E}^p_{ij} \geqq 0, \quad \text{for} \quad \dot{\sigma}_{ij} n_{ij} \geqq 0, \tag{3.19}$$

$$\dot{T}_{ij} \dot{\varepsilon}^p_{ij} \geqq 0, \quad \text{for} \quad \dot{T}_{ij} m_{ij} \geqq 0. \tag{3.20}$$

Suppose the functions f, g are convex. In addition to (3.19) and (3.20) we have the following inequalities

$$(\sigma_{ij} - \sigma^0_{ij}) \dot{E}^p_{ij} \geqq 0, \tag{3.21}$$

$$(\sigma_{ij} - T^0_{ij}) \dot{\varepsilon}^p_{ij} \geqq 0, \tag{3.22}$$

where $\boldsymbol{\sigma}^0$, $\boldsymbol{T}^0$ are arbitrary stresses lying in closed domains determined by $f = 0$, $g = 0$, respectively (Fig. 2).

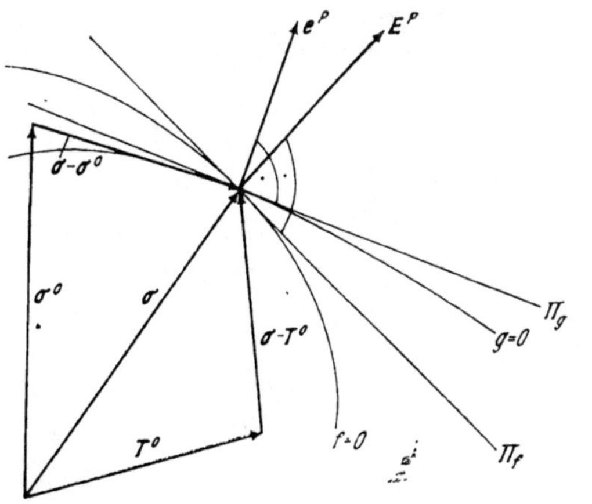

Fig. 2

The set of inequalities (3.19) – (3.22) can be viewed as the counterpart of DRUCKER's postulate for the original and adjoint solids treated as a whole (cf. [27, 29]).

4. Functional expressed in terms of rate of deformations, rate of stresses and velocities

Suppose a linear operator L and thus its adjoint L^* are given. The equivalent variational formulation may then be considered [13, 17]. Such an approach is well known in linear problems of fluid mechanics, heat and mass transfer [17]. The method has also been used to solve some non-conservative problems of stability (cf., for instance [18, 19]).

We shall apply the method of adding the adjoint operator to obtain the variational formulation for the operator L, L^* specified in the previous section.

Before proceeding to the derivation of variational principles, some general considerations are necessary (cf. [13]).

Let on $U_2 \times U_1$ a continuous nondegenerate bilinear form $\langle .,. \rangle$ be prescribed. Suppose that a linear operator $L: U_1 \to U_2$ is not self-adjoint with respect to this form, i.e.

$$\langle Lu_1^1, u_1^2 \rangle = \langle L^* u_1^2, u_1^1 \rangle ; \qquad u_1^1, u_1^2 \in U_1 \tag{4.1}$$

and $L \neq L^*$. Then the linear operator

$$\tilde{L} \overset{\mathrm{df}}{=} \begin{bmatrix} 0 & L^* \\ L & 0 \end{bmatrix} \tag{4.2}$$

is self-adjoint with respect to the bilinear form b, $b: (U_2 \times U_2) \times (U_1 \times U_1) \to R$ defined as follows:

$$b([u_2^1, u_2^2], [u_1^1, u_1^2]) = \langle u_2^1, u_1^1 \rangle + \langle u_2^2, u_1^2 \rangle . \tag{4.3}$$

For the considered equation and its adjoint

$$\left. \begin{array}{l} Lu = G \\ L^* u^* = G^* \end{array} \right\} \Leftrightarrow \begin{bmatrix} 0 & L^* \\ L & 0 \end{bmatrix} \begin{bmatrix} u \\ u^* \end{bmatrix} = \begin{bmatrix} G \\ G^* \end{bmatrix} \tag{4.4}$$

the variational functional has the form

$$F(u, u^*) = \langle Lu, u^* \rangle - \langle G, u^* \rangle - \langle G^*, u \rangle . \tag{4.5}$$

Now we apply the formula (4.5) to derive variational functionals for the considered class of nonstandard materials. Putting $\boldsymbol{u} = (\dot{\boldsymbol{\sigma}}, \boldsymbol{v}, \dot{\boldsymbol{\varepsilon}})$; $\boldsymbol{u}^* = (\dot{\boldsymbol{T}}, \boldsymbol{V}, \dot{\boldsymbol{E}})$ and taking account of (3.1)–(3.5) in (4.5), we obtain

$$\begin{aligned} F(\dot{\boldsymbol{\sigma}}, \boldsymbol{v}, \dot{\boldsymbol{\varepsilon}}; \dot{\boldsymbol{T}}, \boldsymbol{V}, \dot{\boldsymbol{E}}) &= \int_\Omega [(v_{j,i} - \dot{\varepsilon}_{ij}) \dot{T}_{ij} - \dot{\sigma}_{ij,i} V_j - (\dot{\sigma}_{ij} - B_{ijkl} \dot{\varepsilon}_{kl}) \dot{E}_{ij} - B_j v_j - b_j V_j)] \, \mathrm{d}\Omega + \\ &+ \int_{S_1} [-n_i v_j \dot{T}_{ij} + n_i (\bar{V}_j \dot{\sigma}_{ij} + \bar{v}_i \dot{T}_{ij})] \, \mathrm{d}S + \int_{S_2} [n_i V_j \dot{\sigma}_{ij} - P_j v_j - p_j V_j] \, \mathrm{d}S . \end{aligned} \tag{4.6}$$

After some calculations including integration by parts the functional F is shown to be

$$\begin{aligned} F(\dot{\boldsymbol{\sigma}}, \boldsymbol{v}, \dot{\boldsymbol{\varepsilon}}; \dot{\boldsymbol{T}}, \boldsymbol{V}, \dot{\boldsymbol{E}}) &= \int_\Omega [-\dot{\sigma}_{ij} \dot{E}_{ij} - \dot{T}_{ij} \dot{\varepsilon}_{ij} + Y(\dot{\boldsymbol{\varepsilon}}, \dot{\boldsymbol{E}}) + v_{j,i} \dot{T}_{ij} + V_{j,i} \dot{\sigma}_{ij} - B_j v_j - b_j V_j] \, \mathrm{d}\Omega + \\ &+ \int_{S_1} [(-n_i v_j + n_i \bar{v}_j) \dot{T}_{ij} + (-n_i V_j + n_i \bar{V}_j) \dot{\sigma}_{ij}] \, \mathrm{d}S - \int_{S_2} (P_j v_j + p_j V_j) \, \mathrm{d}S \end{aligned} \tag{4.7}$$

where

$$Y(\dot{\boldsymbol{\varepsilon}}, \dot{\boldsymbol{E}}) \overset{\mathrm{df}}{=} B_{ijkl} \dot{\varepsilon}_{kl} \dot{E}_{ij} . \tag{4.8}$$

The functional (4.6) or (4.7) depends on two independent vector fields: $\boldsymbol{v}$, $\boldsymbol{V}$ and four independent tensor fields: $\dot{\boldsymbol{\sigma}}$, $\dot{\boldsymbol{T}}$, $\dot{\boldsymbol{\varepsilon}}$, $\dot{\boldsymbol{E}}$.

Now we can establish the following variational principle: the condition $\delta F = 0$ yields as EULER's equations the relations (3.1–3.5) and (3.9–3.13). The variational equation $\delta F = 0$, where F is given by (4.6) or (4.7), can be viewed as the counterpart of the HU-WASHIZU variational principle (cf. [20, 21]).

5. Functional depending on rate of stresses and velocities

Now we proceed to the elimination of rate of deformations from (4.7). We put

$$W(\dot{\boldsymbol{\sigma}}, \dot{\boldsymbol{T}}) \overset{\mathrm{df}}{=} \dot{\sigma}_{ij} \dot{E}_{ij} + \dot{T}_{ij} \dot{\varepsilon}_{ij} - Y(\dot{\boldsymbol{\varepsilon}}, \dot{\boldsymbol{E}}) , \tag{5.1}$$

where

$$\dot{\varepsilon}_{ij} = A_{ijkl} \dot{\sigma}_{kl} , \qquad \dot{E}_{ij} = A_{klij} \dot{T}_{kl} . \tag{5.2}$$

Substituting (5.2) into (5.1) results in the following expression for the function W:

$$W(\dot{\boldsymbol{\sigma}}, \dot{\boldsymbol{T}}) = A_{ijkl} \dot{\sigma}_{kl} \dot{T}_{ij} . \tag{5.3}$$

Thus we have

$$W + Y = \dot{\sigma}_{ij} \dot{E}_{ij} + \dot{T}_{ij} \dot{\varepsilon}_{ij} . \tag{5.4}$$

The relation (5.4) is the counterpart of the LEGENDRE transformation (cf., for instance [22]). The functions W, Y are thus the counterparts of potentials.

Taking account of (5.4) in (4.7) we obtain the functional

$$\begin{aligned} F_1(\dot{\boldsymbol{\sigma}}, \boldsymbol{v}, \dot{\boldsymbol{T}}, \boldsymbol{V}) &= \int_\Omega [-W(\dot{\boldsymbol{\sigma}}, \dot{\boldsymbol{T}}) + v_{j,i} \dot{T}_{ij} + V_{j,i} \dot{\sigma}_{ij} - B_j v_j - b_j V_j)] \, \mathrm{d}\Omega + \\ &+ \int_{S_1} [(-n_i v_j + n_i \bar{v}_j) \dot{T}_{ij} + (-n_i V_j + n_i \bar{V}_j) \dot{\sigma}_{ij}] \, \mathrm{d}S - \int_{S_2} (P_j v_j + p_j V_j) \, \mathrm{d}S \end{aligned} \tag{5.5}$$

depending on two independent vector fields: $\boldsymbol{v}$, $\boldsymbol{V}$ and two independent tensor fields: $\dot{\boldsymbol{\sigma}}$, $\dot{\boldsymbol{T}}$. The condition $\delta F_1 = 0$ yields as EULER's equations the following relations:

a) for the original solid

$$v_j = \bar{v}_j, \ S_1$$
$$\dot{\sigma}_{ij,i} + b_j = 0, \ \Omega$$
$$n_i\dot{\sigma}_{ij} = p_j, \ S_2$$
$$v_{(j,i)} = A_{ijkl}\dot{\sigma}_{kl}, \ \Omega$$

b) for the adjoint solid

$$V_j = \bar{V}_j, \ S_1$$
$$\dot{T}_{ij,i} + B_j = 0, \ \Omega$$
$$n_i\dot{T}_{ij} = P_j, \ S_2$$
$$V_{(j,i)} = A_{klij}\dot{T}_{kl}, \ \Omega$$

We conclude that the variational equation $\delta F_1 = 0$ is the counterpart of HELLINGER-REISSNER's variational principle (cf., for instance [20, 21]).

6. Functionals expressed in terms of velocities or rate of stresses only

We shall derive two functionals, being much simpler than these derived so far. The first (the second) functional will depend on the velocities (rate of stresses) only.

6.1. Assume the equations of geometrical compatibility (3.1), (3.9) and kinematic boundary conditions (3.2) (3.10) are satisfied. The functional expressed in terms of velocity fields only has the form

$$J(\boldsymbol{v}, \boldsymbol{V}) = F - \left\langle \frac{\partial F}{\partial \dot{\boldsymbol{\sigma}}}, \dot{\boldsymbol{\sigma}} \right\rangle - \left\langle \frac{\partial F}{\partial \dot{\boldsymbol{T}}}, \dot{\boldsymbol{T}} \right\rangle = \int_\Omega [Y(\dot{\boldsymbol{\varepsilon}}, \dot{\boldsymbol{E}}) - B_i v_i - b_i V_i] \, d\Omega - \int_{S_2} (P_i v_i + p_i V_i) \, dS, \tag{6.1}$$

where

$$\dot{\varepsilon}_{ij} = v_{(j,i)}, \quad \dot{E}_{ij} = V_{(j,i)}, \quad \text{and} \quad \frac{\partial F}{\partial \dot{\boldsymbol{\sigma}}}, \frac{\partial F}{\partial \dot{\boldsymbol{T}}}$$

denote partial derivatives (gradients) of the functional F with respect to $\boldsymbol{\sigma}$, $\boldsymbol{T}$ (cf. [23—25]) given by

$$\begin{aligned}
\left(\frac{\partial F}{\partial \dot{\boldsymbol{\sigma}}}\right)_{ij} &= V_{(j,i)} - \dot{E}_{ij}, \ \Omega \\
\left(\frac{\partial F}{\partial \dot{\boldsymbol{\sigma}}}\right)_{ij} &= -n_{(i}V_{j)} + n_{(i}\bar{V}_{j)}, \ S_1 \\
\left(\frac{\partial F}{\partial \dot{\boldsymbol{T}}}\right)_{ij} &= v_{(j,i)} - \dot{\varepsilon}_{ij}, \ \Omega \\
\left(\frac{\partial F}{\partial \dot{\boldsymbol{T}}}\right)_{ij} &= -n_{(i}v_{j)} + n_{(i}\bar{v}_{j)}, \ S_1 .
\end{aligned} \tag{6.2}$$

The formula (6.2) determines components of the respective gradients.

The condition $\delta J = 0$ yields

a) for the original solid

$\dot{\sigma}_{ij,i} + b_j = 0, \ \Omega$

$\delta v_i = 0, \ S_1$ (prescribed boundary condition)

$n_i\dot{\sigma}_{ij} = p_j, \ S_2$ (natural boundary condition)

b) for the adjoint solid

$\dot{T}_{ij,i} + B_j = 0, \ \Omega$

$\delta V_i = 0, \ S_1$ (prescribed boundary condition)

$n_i\dot{T}_{ij} = P_j, \ S_2$ (natural boundary condition)

The variational equation $\delta J = 0$ is the counterpart of the principle of stationary value of the total potential energy cf. [20, 21]).

6.2. Now we derive a functional depending on two fields of rate of stresses only. To this end assume that the equilibrium equations (3.3), (3.11), the boundary conditions (3.4), (3.12) and the flow laws (3.5), (3.13) are fulfilled. The partial derivatives $\frac{\partial F}{\partial \boldsymbol{v}}, \frac{\partial F}{\partial \dot{\boldsymbol{\varepsilon}}}, \frac{\partial F}{\partial \boldsymbol{V}}, \frac{\partial F}{\partial \dot{\boldsymbol{E}}}$ — which will be used in the sequel — are easily shown to be

$$\begin{aligned}
\left(\frac{\partial F}{\partial \boldsymbol{v}}\right)_j &= -\dot{T}_{ij,i} - B_j, \ \Omega & \left(\frac{\partial F}{\partial \boldsymbol{V}}\right)_j &= -\dot{\sigma}_{ij,i} - b_j, \ \Omega \\
\left(\frac{\partial F}{\partial \boldsymbol{v}}\right)_j &= n_i\dot{T}_{ij} - P_j, \ S_2 & \left(\frac{\partial F}{\partial \boldsymbol{V}}\right)_j &= n_i\dot{\sigma}_{ij} - p_j, \ S_2 \\
\left(\frac{\partial F}{\partial \dot{\boldsymbol{\varepsilon}}}\right)_{ij} &= -\dot{T}_{ij} + B_{klij}\dot{E}_{kl}, \ \Omega & \left(\frac{\partial F}{\partial \dot{\boldsymbol{E}}}\right)_{ij} &= -\dot{\sigma}_{ij} + B_{ijkl}\dot{\varepsilon}_{kl}, \ \Omega .
\end{aligned} \tag{6.3}$$

The functional expressed in terms of ratse of stresses has the form

$$J_1(\dot{\sigma}, \dot{T}) = F - \left\langle \frac{\partial F}{\partial v}, v \right\rangle - \left\langle \frac{\partial F}{\partial \dot{\varepsilon}}, \dot{\varepsilon} \right\rangle - \left\langle \frac{\partial F}{\partial V}, V \right\rangle - \left\langle \frac{\partial F}{\partial \dot{E}}, \dot{E} \right\rangle$$

$$= - \int_\Omega W(\dot{\sigma}, \dot{T})\, d\Omega + \int_{S_1} n_i(\bar{V}_j \dot{\sigma}_{ij} + \bar{v}_j \dot{T}_{ij})\, dS\,. \tag{6.4}$$

To derive (6.4) the relation

$$Y(\dot{\varepsilon}, \dot{E}) = B_{ijkl} \dot{\varepsilon}_{kl} \dot{E}_{kl} = \dot{\sigma}_{ij} \dot{E}_{ij} = \dot{\sigma}_{ij} A_{klij} \dot{T}_{kl} = W(\dot{\sigma}, \dot{T}) \tag{6.5}$$

has also been used.

The variational equation $\delta J_1 = 0$ yields

a) for the original solid

$\dot{\varepsilon}_{ij} = v_{(j,i)},\ \Omega$

$v_j = \bar{v}_j,\ S_1$ (the natural boundary condition)

$\delta \dot{\sigma}_{ij,i} = 0,\ \Omega$

$n_i \delta \dot{\sigma}_{ij} = 0,\ S_2$ (the prescribed boundary condition)

b) for the adjoint solid

$\dot{E}_{ij} = V_{(j,i)},\ \Omega$

$V_j = \bar{V}_j,\ S_1$ (the natural boundary condition)

$\delta \dot{T}_{ij,i} = 0,\ \Omega$

$n_i \delta \dot{T}_{ij} = 0,\ S_2$ (the prescribed boundary condition)

The variational equation $\delta J_1 = 0$, can be viewed as the counterpart of the principle of the stationary value of the complementary energy (cf. [20, 21]).

7. The principle of virtual velocities and the principle of virtual changes of rate of stresses for the original and adjoint solids treated as a whole

It is well known that the principle of virtual displacements (velocities) and the principle of virtual changes of stresses (rate of stresses) are of fundamental importance in mechanics (cf., for instance [21, 24, 26, 27]). In this Section we shall extend these principles to the case of the original and adjoint solids treated as a whole. By the way, much confusion exists about terminology concerning such notions as "the principle of virtual work", "the principle of virtual displacements (velocities)" and so on. Clear distinctions are made by HLAVAČEK [28] and we adopt his terminology to the case of rate problems considered in this paper.

7.1. We take two solids, geometrically the same. Let one be subjected to rate of body forces $\boldsymbol{b} = (b_i)$, and boundary conditions like (3.2), (3.4) while the other to $\boldsymbol{B} = (B_i)$ and boundary conditions (3.10), (3.12). Obviously, the original and adjoint solids set an example of such two bodies. We consider the solids in equilibrium, that is rate equilibrium equations are satisfied, under prescribed rate of body forces and boundary conditions.

Calculate the work performed by rate of body forces (b_i), (B_i) and rate of tractions (p_i), (P_i) on virtual velocities (δV_i), (δv_i), respectively,

$$\int_\Omega B_i \delta v_i\, d\Omega + \int_{S_2} P_i \delta v_i\, dS + \int_\Omega b_i \delta V_i\, d\Omega + \int_{S_2} p_i \delta V_i\, dS\,.$$

Here we trest the first (third) and the second (fourth) integrals as the work done by the rate of body forces and rate of tractions acting on the second (first) soild on virtual velocities of the first (second) solid. Using the standard procedure (cf., [21, 27]), we obtain

$$\int_\Omega B_i \delta v_i\, d\Omega + \int_{S_2} P_i \delta v_i\, dS = \int_\Omega \dot{T}_{ij} \delta \dot{\varepsilon}_{ij}\, d\Omega\,, \tag{7.1}$$

$$\int_\Omega b_i \delta V_i\, d\Omega + \int_{S_2} p_i \delta V_i\, dS = \int_\Omega \dot{\sigma}_{ij} \delta \dot{E}_{ij}\, d\Omega \tag{7.2}$$

and eventuallay

$$\int_\Omega (B_i \delta v_i + b_i \delta V_i)\, d\Omega + \int_{S_2} (P_i \delta v_i + p_i \delta V_i)\, dS = \int_\Omega (\dot{\sigma}_{ij} \delta \dot{E}_{ij} + \dot{T}_{ij} \delta \dot{\varepsilon}_{ij})\, d\Omega\,. \tag{7.3}$$

In the case of the original solid and its adjoint the relation (7.3) may be called " the principle of virtual velocities for the original and adjoint solids treated as a whole".

From the formula (7.3) the variational principle $\delta J = 0$ follows. Taking account of (3.5), (3.13) in (7.3) we obtain

$$\delta \left\{ \int_\Omega [B_{ijkl} \dot{\varepsilon}_{kl} \dot{E}_{ij} - B_i v_i - b_i V_i]\, d\Omega - \int_{S_2} (P_i v_i + p_i V_i)\, dS \right\} = 0\,, \tag{7.4}$$

that is the condition of stationary value of the functional (6.1).

7.2. Now we proceed to the derivation of the counterpart of the principle of virtual changes of rate of stresses.

Suppose $\dot{\sigma}$, $\dot{T}$ are actual rate of stresses of the first and the second solid, respectively. We denote by $p_i^{(R)} = n_i \dot{\sigma}_{ij}$, $P_i^{(a)} = n_i \dot{T}_{ij}$, rate of reactions acting on parts S_1 of the first and the second solid, respectively. Consider the expression

$$\int_\Omega v_i\delta B_i \, d\Omega + \int_{S_1} v_i\delta P_i^{(a)} \, dS + \int_{S_2} v_i\delta P_i \, dS + \int_\Omega V_i\delta b_i \, d\Omega + \int_{S_1} V_i\delta p_i^{(R)} \, dS + \int_{S_2} V_i\delta p_i \, dS .$$

Here the first, second and third (the fourth, fifth and sixth) integrals represent the work done by virtual rate of forces, acting on the second (the first) solid, on velocity of the first (the second) body. The standard procedure yields (cf. [21, 27])

$$\int_\Omega v_i\delta B_i \, d\Omega + \int_{S_1} v_i\delta P_i^{(a)} \, dS + \int_{S_2} v_i\delta P_i \, dS = \int_\Omega \dot\varepsilon_{ij}\delta\dot T_{ij} \, d\Omega , \tag{7.5}$$

$$\int_\Omega V_i\delta b_i \, d\Omega + \int_{S_1} V_i\delta p_i^{(R)} \, dS + \int_{S_2} V_i\delta p_i \, dS = \int_\Omega \dot E_{ij}\delta\dot\sigma_{ij} \, d\Omega . \tag{7.6}$$

Thus we have

$$\int_\Omega (v_i\delta B_i + V_i\delta b_i) \, d\Omega + \int_{S_1} v_i\delta P_i^{(a)} + V_i\delta p_i^{(R)}) \, dS + \int_{S_2} (v_i\delta P_i + V_i\delta p_i) \, dS = \int_\Omega (\dot\varepsilon_{ij}\delta\dot T_{ij} + \dot E_{ij}\delta\dot\sigma_{ij}) \, d\Omega . \tag{7.7}$$

In the case of original solid and its adjoint the relation (7.7) may be called "the principle of virtual rate of stresses for the original and adjoint solids treated as a whole".

It should be noted that (7.3) and (7.7) hold irrespective of constitutive laws.

From the formula (7.7) the variational principle $\delta J_1 = 0$ can easily be obtained. To corroborate this statement we put

$$v_i = \bar v_i , \quad V_i = \bar V_i , \quad \text{on } S_1; \quad \delta B_i = \delta b_i = 0 , \quad \delta P_i = \delta p_i = 0 , \quad \dot\varepsilon_{ij} = A_{ijkl}\dot\sigma_{kl} , \quad \dot E_{ji} = A_{klij}\dot T_{kl} ,$$

and obtain

$$\delta \left[\int_\Omega A_{ijkl}\dot\sigma_{kl}\dot T_{ij} \, d\Omega - \int_{S_1} n_i(\bar V_j\dot\sigma_{ij} + \bar v_i\dot T_{ij}) \, dS \right] = 0 \tag{7.8}$$

that is the condition of the stationary value of the functional (6.4).

8. Reciprocal theorems

In this section we drive two reciprocal theorems valid for elastic-plastic solids, obeying non-associated flow laws.

8.1. Let u, u^* be a solution of the system (4.4). Multiplying $(4.4)_1$ by u^* while $(4.4)_2$ by u we obtain

$$\langle Lu, u^*\rangle = \langle G, u^*\rangle , \qquad \langle L^*u^*, u\rangle = \langle G^*, u\rangle .$$

Subtracting we arrive at (cf. [30, 31] in the case of HILBERT spaces)

$$\langle G, u^*\rangle = \langle G^*, u\rangle . \tag{8.1}$$

Applying the reciprocal formula (8.1) to the considered problem of non-associated elasto-plasticity we obtain

$$\int_\Omega b_iV_i \, d\Omega - \int_{S_1} n_j\bar v_i\dot T_{ij} \, dS + \int_{S_2} p_iV_i \, dS = \int_\Omega B_iv_i \, d\Omega - \int_{S_1} n_j\bar V_i\dot\sigma_{ij} \, dS + \int_{S_2} P_iv_i \, dS . \tag{8.2}$$

8.2. DE BOER [14] has formulated and proved a reciprocal theorem valid for elastic-plastic materials with hardening and described by associated flow laws. In the case of associated plasticity a major role is played by the symmetry of A_{ijkl}, or B_{ijkl}, with respect to the group of indices ij and kl. As we know, non-associated flow laws lead to $A_{ijkl} \neq A_{klij}$. We conclude that the reciprocal theorem of DE BOER does not hold for solids with non-associated flow laws. Generally speaking, it is due to the non-self-adjointness of L. On the other hand the operator $\tilde L$, given by (4.2) is self-adjoint. Thus it should be possible to extend the reciprocal theorem of DE BOER to the case of the original and adjoint solids treated as a whole. We shall prove this statement.

In the sequel, for the sake of simplicity, we assume $\bar v_i = \bar V_i = 0$ on S_1. We take the system (4.4) with $G = G_\alpha$, $G^* = G^*_\alpha$, $\alpha = 1, 2$, that is

$$\begin{cases} Lu_1 = G_1 \\ L^*u_1^* = G_1^* \end{cases} \qquad \begin{cases} Lu_2 = G_2 \\ L^*u_2^* = G_2^* . \end{cases}$$

Multiplying and adding we arrive at

$$\langle Lu_1, u_2^*\rangle + \langle L^*u_1^*, u_2\rangle = \langle G_1, u_2^*\rangle + \langle G_1^*, u_2\rangle , \tag{8.3}$$

$$\langle Lu_2, u_1^*\rangle + \langle L^*u_2^*, u_1\rangle = \langle G_2, u_1^*\rangle + \langle G_2^*, u_1\rangle . \tag{8.4}$$

Because L^* is the adjoint of L thus the right-hand sides of (8.3) and (8.4) are equal, i.e.

$$\langle G_1, u_2^*\rangle + \langle G_1^*, u_2\rangle = \langle G_2, u_1^*\rangle + \langle G_2^*, u_1\rangle . \tag{8.5}$$

Now we apply the reciprocal theorem (8.5). Take two sets of rate of body forces: $(b_i^{(\alpha)})$, $(B_i^{(\alpha)})$ and rate of surface tractions $(p_i^{(\alpha)})$, $(P_i^{(\alpha)})$ producing in the original and adjoint solids rate of stresses $(\dot\sigma_{ij}^{(\alpha)})$, $(\dot T_{ij}^{(\alpha)})$ and velocities $(v_i^{(\alpha)})$, $(V_i^{(\alpha)})$. The reciprocal theorem (8.5) yields

$$\int_\Omega (B_i^{(1)}v_i^{(2)} + b_i^{(1)}V_i^{(2)})\,d\Omega + \int_{S_2} (P_i^{(1)}v_i^{(2)} + p_i^{(1)}V_i^{(2)})\,dS = \int_\Omega (B_i^{(2)}v_i^{(1)} + b_i^{(2)}V_i^{(1)})\,d\Omega + \int_{S_2} (P_i^{(2)}v_i^{(1)} + p_i^{(2)}V_i^{(1)})\,dS\,. \tag{8.6}$$

The formula (8.6) may also be derived by the procedure similar to that used in [14]. For the purpose the following relation, derived in an analogous way as the principle (7.3), is fundamental

$$\int_\Omega (B_iv_i + b_iV_i)\,d\Omega + \int_{S_2} (P_iv_i + p_iV_i)\,dS = \int_\Omega (\sigma_{ij}\dot E_{ij} + \dot T_{ij}\dot\varepsilon_{ij})\,d\Omega\,. \tag{8.7}$$

Here (v_i), (V_i), are kinematically admissible that is fulfilling strain-displacement relations and $v_i = V_i = 0$ on S_1, while (σ_{ij}), $(\dot T_{ij})$ are statically admissible, thus satisfying rate-equilibrium equations and boundary conditions on S_2. The formula (8.7) may be called "the principle of virtual work for the original and adjoint solids treated as a whole". (cf. [28]). The principle (8.7) allows to prove (8.6) by the reasoning similar to that of DE BOER [14].

Reciprocal theorems, useful in calculation of displacements due to inelastic part of the strain tensor, have been derived in [4, 32] (cf. also [33]).

9. Conclusions and final remarks

In the paper [31] we have derived dual extremum principles for a similar class of elastic-plastic nonstandard materials, but under the assumption of positive-definiteness of $\boldsymbol{B} = (B_{ijkl})$. The assumption like this puts often too severe restrictions on the behaviour of materials [2, 9]. For instance, it may even rule out softening. Thus only variational principle are available in general.

Variational principles for rigid-plastic solids obeying non-associated flow laws have been derived in [34, 35]. This case is still more complicated because the boundary value operator is nonlinear.

The method of adding the adjoint operator has enabled to establish counterparts of the principle of stationary value of potential energy and the principle of stationary value of complementary energy. The variational principles corresponding to the principles of HU-WASHIZU and HELLINGER-REISSNER has also been formulated. The derived variational principles rank among very important approximative methods, cf. [36].

Treating the original and adjoint solids as a whole, the counterparts of the principle of virtual velocities, the principles of virtual changes of rate of stresses and the principle of virtual work have been formulated. We think that similar principles can turn out to be of fundamental importance in non-associated plasticity in general, and not only for solids briefly described in Section 2.

Discontinuities in velocities and/or rate of stresses can be included similarly as in [24, 25, 35].

It should be noted that the variational principles derived in this paper, hold both for hardening and/or softening behavior and coupling between elastic and plastic deformations irrespective of the convexity of a yield function and plastic potential.

Acknowledgements. The idea of using the method of adding the adjoint operator was stimulaed by discussions with Professor Z. MRÓZ of Polish Academy of Sciences to whom we are greatly indebted. We acknowledge also the reading of the manuscript by Professor M. J. SEWELL of Reading University, England.

References

1 MRÓZ, Z., Non-associated flow laws in plasticity, J. de Méc., 2, 21—42 (1963).
2 MRÓZ, Z., On forms of constitutive laws for elastic-plastic solids, Arch. Mech. Stos., 18, 3—35 (1966).
3 MELAN, E., Zur Plastizität des räumlichen Kontinuums, Ing.-Arch., 9, 116—126 (1938).
4 MAIER, G., A minimum principle for incremental elasto-plasticity with non-associated flow laws, J. Mech. Phys. Solids, 18, 319—330 (1970).
5 NGUYEN, Q. S., BUI, H. D., Sur les matériaux élastoplastiques à écrouissage positif ou négatif, J. de Méc., 13, 321—342 (1974).
6 HUECKEL, T., On the description of nonlinear elasticity of granular media, Mech. Teor. i Stos., 14, 335—345 (1976) (in Polish).
7 HUECKEL, T., Coupling of elastic and plastic deformations of bulk solids, Meccanica, 11, 227—235 (1976).
8 HUECKEL, T.; MAIER, G., Incremental boundary value problems in the presence of coupling of elastic and plastic deformations: a rock mechanics oriented theory, Int. J. Solids Struct., 13, 1—15 (1977).
9 MAIER, G., HUECKEL, T., Nonassociated and coupled flow rules of elastoplasticity for geotechnical media, 9-th Int. Conf. of the Int. Soc. for Soil Mech. and Found. Eng., Tokyo, July 1977.
10 WONG, P. K. K.; MITCHELL, R. J., Yielding and plastic flow of sensitive cemented clay, Géotechnique, 25, 763—782 (1975).
11 TRĄMPCZYŃSKI, W., Experimental study of kinematics of soil shoving, Bull. Acad. Pol. Sci., Série Sci. Techn., 25, 985—993 (1977).
12 TELEGA, J. J., On plane plastic flow of compressible solids, ZAMM, 58, 133—142 (1978).
13 TONTI, E., On the variational formulation for linear initial value problems, Annali di Mat. Pura ed Appl., 95, 331—359 (1973).
14 DE BOER, R., Das Reziprozitätstheorem in der Plastizitätstheorie, ZAMP, 26, 633—636 (1975).
15 ADAMS, R., Sobolev spaces, Academic Press, New York, 1975.
16 MAGRI, F., Variational formulation for every linear equation, Int. J. Eng. Sci., 12, 537—549 (1974).
17 FINLAYSON, B. A., The method of weighted residuals and variational principles with application in fluid mechanics heat and mass transfer, Academic Press, New York and London, 1972.
18 ANDERSON, G. L., On the stability of a deep beam subjected to non-conservative and dissipative forces, Meccanica, 9, 102—110 (1974).

19 Leipholz, H. H. E., On the calculation of buckling loads by means of hybrid Ritz equations, Arch. of Mech., **25**, 895—901 (1973).
19a Leipholz, H. H. E., Direct variational methods and eigenvalue problems in engineering, Noordhoff International Publishing, Leyden, 1977.
20 Hlaváček, I., Derivation of nonclassical variational principles in the theory of elasticity, Aplikac. Mat., **12**, 15—29 (1967).
21 Washizu, K., Variational methods in elasticity and plasticity, Pergamon Press, Oxford et al., 1968.
22 Olszak, W.; Perzyna, P., Remarks on the validity of variational theorems in the mechanics of inelastic anisotropic bodies, in "Non-homogeneity in elasticity and plasticity", Proc. I.U.T.A.M., Symp., Warsaw 1958, 157—165.
23 Vainberg, M. M., Variacionnye metody issledovanija nelinejnych operatorov (Variational methods in the study on nonlinear operators), Gos. Izd. Tech. — Teor. Lit., Moskva, 1956.
24 Sewell, M. J., The governing equations and extremum principles of elasticity and plasticity generated from a single functional, Part 1, J. Struct. Mech., **2**, 1—32 (1973).
25 Telega, J. J., An operator approach to the formulation of variational principles for plastic solids, Mech. Teor. i Stos., **14**, 557 to 570 (1976) (in Polish).
26 Lippmann, H., Extremum and variational principles in mechanics, CISM Coures and Lectures — No. 54, Springer Verlag Wien-New York 1972.
27 Fung, Y. C., Foundations of solid mechanics, Prentice-Hall, New Jersey, 1965.
28 Hlaváček, I., Variational principles in the linear theory of elasticity for general boundary conditions, Apl. Mat., **12**, 425—448 (1967).
29 Olszak, W.; Perzyna, P.; Sawczuk, A., (editors) Teoria plastyczności (The theory of plasticity), PWN, Warszawa, 1965 (in Polish).
30 Collins, W. D., Dual extremum principles for the heat equation, Proc. Roy. Soc. Edinb., **77A**, 273—292 (1977).
31 Telega, J. J., Dual extremum principles in rate boundary value problems of non-associated plasticity, Int. J. Eng. Sci., **17**, 215—226 (1979)
32 Lin, T. H., Reciprocal theorem for displacements in inelastic bodies J. Comp. Mater., **1**, 144—151 (1967).
33 Dorosz, S., An upper bound to maximum residual deflections of elastic-plastic structures at shakedown, Bull. Acad. Pol. Sci., Série Sci. Techn., **24**, 167—174 (1976).
34 Telega, J. J., Variational principles for an incompressible, perfectly plastic material with non-associated flow law, Bull. Acad. Pol. Sci., Série Sci. Techn. **25**, 977—983 (1977).
35 Telega, J. J., Derivation of variational principles for rigid-plastic solids obeying non-associated flow laws, (submitted).
36 Telega, J. J., The finite element method in the mechanics of deformable bodies, IFTR Reports, 55 (1975) (in Polish).

Eingereicht am 10. 11. 1977, in revidierter Fassung am 26. 9. 1978

Anschrift: Dr. Józef Joachim Telega, Institute of Fundamental Technological Research, Polish Academy of Sciences, Department of Structural Mechanics, ul. Świetokrzyska 21, 00049 Warsaw, Poland

ZAMM 60, 83 – 88 (1980)

N. GASS

Solution of Einstein's Gravitational Field Equations by the Finite Element Method

Einsteins Gravitationsgleichungen im materie-erfüllten Raum sind in Form eines diskretisierten Funktionals gegeben, aus welchem nach Durchführung eines Extremalprozesses ein System von nichtlinearen algebraischen Gleichungen entsteht. Auf dieser Basis wird ein 4-dimensionales finites Element entwickelt. Für den Fall *sphärischer Symmetrie und Zeitunabhängigkeit wird ein numerisches Beispiel gegeben.*

Einstein's field equations of gravitation in presence of matter are presented in the form of a discrete functional from which, after performing an extremization process, a system of nonlinear algebraic equations arises. On this basis, a 4-dimensional finite element is developed and in the case of spherical symmetry and time-independence, a numerical example is carried out.

Уравнения Эйнштейна для гравитации при отсутствия материи представляются в виде дискретного функционала, из которого, после проведения некоторого процесса экстремизация, получается система нелинейных алгебраических уравнений. На основе этого конструируется четырехмерный финитный элемент и в случае сферической симметрии и независимости от времени даестя численный пример.

1. Introduction

The gravitational field equations of EINSTEIN do not lend themiselves to easy investigation. The reason for this lies on one hand in the nonlinearity of these equations and on the other hand in the large number of variables needed to describe the metric of space-time. There is no general classical method of finding rigorous solutions of the field equations. However, the equations have been solved in a few cases in which the number of variables were reduced by symmetry conditions [1]. Another route to solve the gravitational field equations is by means of numerical methods based on variational principles. Amongst all, the finite element method [2] has already been proven successful in the solution of complex equation in physics [3, 4] and nonlinear field-equations in mechanics [5].

In the present paper now a *four-dimensional finite element* is developed and in the special case of spherical symmetry in space and time-independence, a numerical example is given to demonstrate the present procedure.

2. The gravitational field equations and their variational representation

EINSTEIN's field equations of gravitation in presence of matter [6]

$$G_{\mu\nu} = R_{\mu\nu} - \tfrac{1}{2}\, g_{\mu\nu} R = -8\pi k T_{\mu\nu} \tag{1}$$

are the direct generalizations of NEWTON's field equation

$$\nabla^2 G = -4\pi\varrho \,.$$

Equation (1) describes the gravitational field in terms of the metric tensor $g_{\mu\nu}$ (or $g^{\mu\nu}$) whose components assume the role of gravitational potentials. $R_{\mu\nu}$ denotes the RICCI tensor, k is the gravitational constant and $T_{\mu\nu}$ represents the stress-energy tensor of a given matter distribution. Clearly, if gravitational matter is concentrated in small regions (mass points) and zero elsewhere then the field equations reduce to $G_{\mu\nu} = 0$.

Equation (1) may also be expressed in the equivalent variational form [6] as following

$$J = \int\limits_{\mathcal{D}} (R + 8\pi k \mathcal{M}) \sqrt{-g}\, \mathrm{d}\tau \,, \tag{2}$$

where the gravitational part of $\mathcal{M}$ can be written as

$$\frac{\partial(\sqrt{-g}\, M)}{\partial(\sqrt{-g}\, g^{\mu\nu})} = -T_{\mu\nu} \tag{3}$$

and

$$g = \det\,(g_{\mu\nu}) \,.$$

A set of coordinates may be introduced such that $\sqrt{-g} = 1$ which reduces (1) to $G_{\mu\nu} = R_{\mu\nu}$. Further, on the assumption of uniform distributed matter in space and time, the stress-energy tensor is independent of $g_{\mu\nu}$ and equation (3) may be integrated to obtain

$$M = -T_{\mu\nu} g^{\mu\nu} \,. \tag{4}$$

Now, in the finite element method the continuum $\mathcal{D}$ is fictitiously cut into p smaller subregions $\mathcal{D}_p$ where the surface of such an element consists of 2 mutually exclusive parts, i.e.

$$S_p = S_{op} + S_{gp} \,.$$

On S_{gp} the *generalized gravitational potentials* $\bar{P}_{\mu\nu}$ are prescribed while on $\grave{S}_{op}$ nothing is prescribed. With the previous choice of $\sqrt{-g} = 1$ and $R = g^{\mu\nu} R_{\mu\nu}$ the functional J may be postulated as

$$J = \int_{\mathfrak{D}_p} \sum_p (g^{\mu\nu} \Gamma^{\alpha}_{\mu\beta} \Gamma^{\beta}_{\nu\alpha} - 8\pi k T_{\mu\nu} g^{\mu\nu}) \, d\tau - \int_{S_{gp}} \sum_p g^{\mu\nu} \bar{P}_{\mu\nu} \, dS , \tag{5}$$

where the $\Gamma^{\alpha}_{\beta\nu}$ are the CHRISTOFFEL symbols of the second kind.

Theorem: *Amongst all the admissible potentials $g^{\alpha\beta}$, the true solution is distinguished by the stationary conditions of the functional J.*

To prove the theorem, the variation of J is equated to zero and thence the conditions deduced associated with the true solution. Taking the variation of the first term in equation (5) yields

$$\delta(g^{\mu\nu} \Gamma^{\alpha}_{\mu\beta} \Gamma^{\beta}_{\nu\alpha}) = \Gamma^{\alpha}_{\mu\beta} \Gamma^{\beta}_{\nu\alpha} \, \delta g^{\mu\nu} + 2 g^{\mu\nu} \Gamma^{\alpha}_{\mu\beta} \, \delta \Gamma^{\beta}_{\nu\alpha} = -\Gamma^{\alpha}_{\mu\beta} \Gamma^{\beta}_{\nu\alpha} \, \delta g^{\mu\nu} + \Gamma^{\alpha}_{\mu\beta} \, \delta g^{\mu\beta}_{,\alpha} .$$

Invoking the integral theorem of GAUSS-OSTROGRADSKI

$$\int_V X_\beta \, \delta Y^{\alpha\beta}_{,\alpha} \, dV = \int_S X_\beta n_\alpha \, \delta Y^{\alpha\beta} \, dS - \int_V X_{\beta,\alpha} \, \delta Y^{\alpha\beta} \, dV$$

the last term in above expression may be modified accordingly and the variation of J becomes

$$\delta J = \int_{\mathfrak{D}_p} \sum_p (-\Gamma^{\alpha}_{\mu\beta} \Gamma^{\beta}_{\nu\alpha} - \Gamma^{\beta}_{\mu\nu,\alpha} - 8\pi k T_{\mu\nu}) \, \delta g^{\mu\nu} \, d\tau + \int_{S_{gp}} \sum_p (\Gamma^{\alpha}_{\mu\nu} n_\alpha - \bar{P}_{\mu\nu}) \, \delta g^{\mu\nu} \, dS . \tag{6}$$

Now since the $\delta g^{\mu\nu}$ may assume arbitrary values, the vanishing of δJ requires the conditions

$$\sum_p (T^{\alpha}_{\mu\nu} n_\alpha - \bar{P}_{\mu\nu}) = 0 , \tag{7}$$

$$\sum_p (\Gamma^{\alpha}_{\mu\nu,\alpha} + \Gamma^{\alpha}_{\mu\beta} \Gamma^{\beta}_{\nu\alpha}) = -\sum_p 8\pi k T_{\mu\nu} . \tag{8}$$

Equations (8) represent the *gravitational field equation in the case of* $\sqrt{-g} = 1$ [7] and equations (7) are the *associated boundary conditions.*

The implementation of the CHRISTOFFEL symbols

$$\Gamma^{\beta}_{\mu\alpha} = \tfrac{1}{2} g^{\beta\nu} (g_{\nu\mu,\alpha} + g_{\nu\alpha,\mu} - g_{\mu\alpha,\nu}) \tag{9}$$

in equation (5) yields the functional for one single element as

$$J_e = \int_{\mathfrak{D}_p} H \, d\tau - \int_{S_{gp}} g^{\mu\nu} \bar{P}_{\mu\nu} \, dS , \tag{10}$$

where the HAMILTONian is defined by

$$H = \tfrac{1}{2} g_{\alpha\beta} g^{\alpha\mu}_{,\nu} g^{\beta\nu}_{,\mu} - \tfrac{1}{4} g^{\alpha\beta} g_{\mu\nu} g_{\gamma\varrho} g^{\mu\gamma}_{,\alpha} g^{\nu\varrho}_{,\beta} - 8\pi k T_{\mu\nu} g^{\mu\nu} . \tag{11}$$

At this stage, EINSTEIN's *weak principle of equivalence* is introduced with the consequence that for weak and slow changing fields the metric tensor may be defined as [6]

$$g_{\alpha\beta} = \varepsilon_{\alpha\beta} + h_{\alpha\beta} , \qquad g^{\alpha\beta} = \varepsilon^{\alpha\beta} + h^{\alpha\beta} \tag{12}$$

where $\varepsilon_{\alpha\beta} = \varepsilon^{\alpha\beta} = \begin{pmatrix} -1 & & & 0 \\ & -1 & & \\ & & -1 & \\ 0 & & & +1 \end{pmatrix}$ is the LORENTZ metric and $h_{\alpha\beta}$ is a first order quantity with the transformation law

$$h_{\mu\nu} = -\varepsilon_{\mu\alpha} \varepsilon_{\nu\beta} h^{\alpha\beta} . \tag{13}$$

The determinant is given by

$$g = -(1 + \varepsilon^{\alpha\beta} h_{\alpha\beta}) . \tag{14}$$

Substituting equation (12) into (11) yields finally the HAMILTON *function* as

$$\begin{aligned} H = {} & \tfrac{1}{2} \varepsilon_{\alpha\alpha} h^{\alpha\beta}_{,\gamma} h^{\alpha\gamma}_{,\beta} - \tfrac{1}{4} \varepsilon^{\alpha\alpha} \varepsilon_{\beta\beta} \varepsilon_{\gamma\gamma} h^{\beta\gamma}_{,\alpha} h^{\beta\gamma}_{,\alpha} - \tfrac{1}{2} \varepsilon_{\alpha\alpha} \varepsilon_{\beta\beta} h^{\alpha\beta} h^{\alpha\gamma}_{,\varrho} h^{\beta\gamma}_{,\gamma} - \tfrac{1}{4} \varepsilon_{\alpha\alpha} \varepsilon_{\beta\beta} h^{\beta\beta} h^{\alpha\varrho}_{,\mu} h^{\alpha\mu}_{,\varrho} - \tfrac{1}{4} \varepsilon_{\alpha\alpha} \varepsilon_{\beta\beta} h^{\gamma\gamma} h^{\alpha\beta}_{,\gamma} h^{\alpha\beta}_{,\gamma} + \\ & + \tfrac{1}{2} \varepsilon^{\alpha\alpha} \varepsilon_{\beta\beta} \varepsilon_{\gamma\gamma} \varepsilon_{\gamma\gamma} h^{\gamma\beta}_{,\alpha} h^{\gamma\beta}_{,\alpha} h^{\gamma\gamma} + \tfrac{1}{8} \varepsilon^{\alpha\alpha} \varepsilon_{\beta\beta} \varepsilon_{\gamma\gamma} \varepsilon_{\varrho\varrho} h^{\varrho\varrho} h^{\beta\gamma}_{,\alpha} h^{\beta\gamma}_{,\alpha} + 8\pi k T_{\alpha\alpha} (\varepsilon^{\alpha\alpha} + h^{\alpha\alpha}) . \end{aligned} \tag{15}$$

3. The finite element formulation

We now construct a discrete model which approximates in its domain the gravitational field equations piecewise over a finite number of subdomains, the finite elements. As main variables we choose the $h^{\alpha\beta}$ according to the transformation

$$h^{\alpha\beta} = \begin{cases} h^{\alpha\alpha} & \text{for} \quad \alpha = \beta \\ 0 & \text{for} \quad \alpha \neq \beta . \end{cases}$$

It should be noted that only first order derivatives appear in the HAMILTON function (15) and hence no derivatives of $h^{\alpha\alpha}$ as nodal quantities are necessary. Thus, we make the assumption

$$h^{\alpha\alpha}(\boldsymbol{x}) = \Psi^{\alpha}_{n}(\boldsymbol{x}) \, G^n , \qquad n = 1, N , \quad N = \text{number of nodes per element} \tag{16}$$

where Ψ are the known shape functions of a particular element and G is the vector of the unknown nodal quantities.

The *gravitational boundary tractions* may be either of NEUMANN type

$$\bar{P}_{\alpha\alpha} = \frac{\partial h_{\alpha\alpha}}{\partial n_\alpha}\bigg|_{\mathscr{B}} \tag{17}$$

or of DIRICHLET type

$$\bar{P}_{\alpha\alpha} = h_{\alpha\alpha}|_{\mathscr{B}} \tag{18}$$

and as such P does not participate in the process of variation. Now the substitution of the shape functions (16) into equation (10) yields the discretized functional for a single element

$$4J_e = E_{mn}G^mG^n - H_{mnk}G^mG^nG^k - 2Q_mG^m + 4\pi kS_mG^m + C\,, \tag{19}$$

where the *element matrices* are given as

$$E_{mn} = \int_{\mathscr{D}_e} [2\varepsilon_{\alpha\alpha}\Psi^\alpha_{m,\alpha}\Psi^\alpha_{n,\alpha} - \varepsilon^{\alpha\alpha}\varepsilon_{\beta\beta}\varepsilon_{\beta\beta}\Psi^\beta_{m,\alpha}\Psi^\beta_{n,\dot{\alpha}}]\,\mathrm{d}\tau\,,$$

$$H_{mnk} = \int_{\mathscr{D}_e} [2\varepsilon^{\alpha\alpha}\varepsilon_{\beta\beta}\Psi^\beta_{m,\alpha}\Psi^\beta_{n,\alpha}\Psi^\beta_k + \tfrac{1}{2}\,\varepsilon^{\alpha\alpha}\varepsilon_{\beta\beta}\varepsilon_{\beta\beta}\varepsilon_{\gamma\gamma}\Psi^\beta_{m,\alpha}\Psi^\beta_{n,\alpha}\Psi^\gamma_k - 2\varepsilon_{\alpha\alpha}\varepsilon_{\alpha\alpha}\Psi^\alpha_{m,\alpha}\Psi^\alpha_{n,\alpha}\Psi_k - \varepsilon_{\alpha\alpha}\varepsilon_{\beta\beta}\Psi^\alpha_{m,\alpha}\Psi_{n,\alpha}\Psi^\beta_k$$

$$- \varepsilon_{\alpha\alpha}\varepsilon_{\alpha\alpha}\Psi^\alpha_{m,\beta}\Psi^\alpha_{n,\beta}\Psi^\beta_k]\,\mathrm{d}\tau\,, \tag{20}$$

$$Q_m = \int_{S_e} \Psi^\alpha_m P_\alpha\,\mathrm{d}S\,, \qquad S_m = 8\int_{\mathscr{D}_e} \Psi^\alpha_m T_{\alpha\alpha}\,\mathrm{d}\tau\,, \qquad C = 32\pi k\int_{S_e} T_{\alpha\alpha}\varepsilon^{\alpha\alpha}\,\mathrm{d}\tau\,.$$

Now, the assemblage of the elements follows the well known procedure. The elements are assembled by transforming the element matrices from the local into the global coordinate system. The transformation matrix may conveniently be expressed by

$$t^{ij} = \delta^{i1}\delta^{jI} + \delta^{i2}\delta^{jJ} + \delta^{i3}\delta^{jK} + \delta^{i4}\delta^{jL} + \dots\,, \qquad i = 1, N;\ j = 1, M,\ M = \text{total number of nodes}\,, \tag{21}$$

where δ_{mn} is the KRONECKER symbol and $I, J, K, L, \dots$ denote the global indices of the nodes of one element. Hence,

$$G^m = \sum_n^N t^{mn}G^n\,, \qquad \boldsymbol{S}_n = \sum_n^N t^{mn}S_m\,, \qquad \boldsymbol{Q}_n = \sum_n^N t^{mn}Q_m\,, \qquad \boldsymbol{C}_n = \sum_n^N t^{mn}C_m\,,$$

$$\boldsymbol{E}_{mn} = \sum_i^N\sum_j^N t^{im}t^{jn}E_{ij}\,, \qquad \boldsymbol{H}_{mnk} = \sum_i^N\sum_j^N\sum_l^N t^{im}t^{jn}t^{lk}H_{ijl}\,. \tag{22}$$

The *functional of the total system* becomes

$$4J = \boldsymbol{E}_{mn}\boldsymbol{G}^m\boldsymbol{G}^n - \boldsymbol{H}_{mnk}\boldsymbol{G}^m\boldsymbol{G}^n\boldsymbol{G}^k - 2\boldsymbol{Q}_m\boldsymbol{G}^m + 4\pi k\boldsymbol{S}_m\boldsymbol{G}^m + \boldsymbol{C} \tag{23}$$

and after performing the variation $\delta J = 0$ with respect to $\boldsymbol{G}$, the *system equations* emerge as

$$2\boldsymbol{E}_{mn}\boldsymbol{G}^n - 3\boldsymbol{H}_{mnk}\boldsymbol{G}^n\boldsymbol{G}^k = 2\boldsymbol{Q}_m - 4\pi k\boldsymbol{S}_m\,. \tag{24}$$

Clearly, the linear equation $\boldsymbol{E}_{mn}\boldsymbol{G}^n = -4\pi k\boldsymbol{S}_m$ corresponds to NEWTON's potential equation of POISSON type for an infinite space ($\boldsymbol{Q} = 0$) with a matter density distribution $\boldsymbol{S} = \varrho$. For the empty space with a single concentrated mass, the equation $\boldsymbol{E}_{mn}\boldsymbol{G}^n = \boldsymbol{Q}_m$, describing the potential field outside the mass point ($\boldsymbol{S}_m = 0$), corresponds to NEWTON's potential equation of LAPLACE type with the well-known solution $G = -m/r$. The nonlinear term $\boldsymbol{H}_{mnk}\boldsymbol{G}^n\boldsymbol{G}^k$ in equation (24) accounts for the deviation from NEWTON's law.

Also it is worth noting that, in this term, the $g^{\alpha\alpha}$ are coupled and hence equation (24) must be solved for all $g^{\alpha\alpha}$ simultaneously. This is the major difference to the linear approach. $\boldsymbol{Q}$ represents the boundary reaction. For the DIRICHLET boundary condition, $\boldsymbol{Q}$ may conveniently be determined by partitioning the system matrix

$$\begin{bmatrix} \boldsymbol{E}_{AA} & \boldsymbol{E}_{AB} \\ \boldsymbol{E}^t_{AB} & \boldsymbol{E}_{BB} \end{bmatrix} \begin{bmatrix} \boldsymbol{G}_A \\ \boldsymbol{G}_B \end{bmatrix} = \begin{bmatrix} \boldsymbol{Q} \\ 0 \end{bmatrix} \tag{25}$$

where $\boldsymbol{G}_A$ is the non-zero vector of the known boundary potentials and $\boldsymbol{G}_B$ is the vector of the unknown potentials inside the domain. The above system may be resolved to give

$$\boldsymbol{E}_{AA}\boldsymbol{G}_A + \boldsymbol{E}_{AB}\boldsymbol{G}_B = \boldsymbol{Q}\,, \qquad \boldsymbol{E}^t_{AB}\boldsymbol{G}_A + \boldsymbol{E}_{BB}\boldsymbol{G}_B = 0\,. \tag{26}$$

From the latter equation the unknown potentials may be calculated as

$$\boldsymbol{G}_B = -\boldsymbol{E}^{-1}_{BB}\boldsymbol{E}^t_{AB}\boldsymbol{G}_A \tag{27}$$

and back-substitution into the first equation yields the *boundary reaction* as

$$\boldsymbol{Q} = \boldsymbol{E}_{AA}\boldsymbol{G}_A - \boldsymbol{E}_{AB}\boldsymbol{E}^{-1}_{BB}\boldsymbol{E}^t_{AB}\boldsymbol{G}_A\,. \tag{28}$$

As boundary potentials $\boldsymbol{G}_A$, NEWTON's values may be employed.

The boundary reaction Q of an element, for the NEUMANN type of boundary condition, may be determined by assuming the following conditions

$$P_\alpha = \frac{\partial h_{\alpha\alpha}}{\partial n} = \varepsilon^{\alpha\alpha}\varepsilon^{\beta\beta}\cdot\cos(n, x^\beta)\,\Phi_{\alpha,\beta}, \tag{29}$$

where Φ denotes a known boundary function (e.g. NEWTON's potential). Hence the term in equation (20) becomes

$$Q_m = \oint_{S_e} \Psi^\alpha \cos(n, x^\beta)\, \varepsilon^{\alpha\alpha}\varepsilon^{\beta\beta}\Phi_{\alpha,\beta}\, dS\,. \tag{30}$$

Invoking the formula of GAUSS-OSTROGRADSKI,

$$\int_\Omega \varphi^\alpha_{,\beta}\, d\Omega = \oint_S \varphi^\alpha \cos(n, x^\beta)\, dS \tag{31}$$

permits the transformation of the boundary integral (30) into the volume integral

$$Q_m = \int_{\mathfrak{D}_e} \varepsilon^{\alpha\alpha}\varepsilon^{\beta\beta}\, \partial/\partial x^\beta\, (\Psi^\alpha_m \Phi_{\alpha,\beta})\, d\tau\,. \tag{32}$$

Now the solution of the system equations may be carried out in a number of ways. For sake of completeness we briefly outline the NEWTON-RAPHSON procedure which, in this particular case, assures fast convergence due to the weak nonlinear influence. Denoting the m-th equation of the set (24) by

$$F_m = 2\boldsymbol{E}_{mn}\boldsymbol{G}^n - 3\boldsymbol{H}_{mnk}\boldsymbol{G}^n\boldsymbol{G}^k - 2\boldsymbol{Q}_m - 4\pi k\boldsymbol{S}_m \tag{33}$$

and letting $\boldsymbol{G}^n_0$ be an approximation for the correct solution, then by TAYLOR series expansion it follows that

$$F_m(\boldsymbol{G}_0) + \sum_{j=1}^{M} \frac{\partial F_m}{\partial \boldsymbol{G}^i}\, \Delta \boldsymbol{G}^j = 0\,, \tag{34}$$

where the gradient of F_m is readily obtained from equation (33) as

$$\frac{\partial F_m}{\partial \boldsymbol{G}^i} = \boldsymbol{E}_{mi} - 3\boldsymbol{H}_{mni}\boldsymbol{G}^n_0\,. \tag{35}$$

The *advantage* of this algorithm is that the vector $\Delta\boldsymbol{G}$ is determined by solving the linear system (34). The next better approximation for the true solution is then given by

$$\boldsymbol{G}^i = \boldsymbol{G}^i_0 + \Delta \boldsymbol{G}^i\,. \tag{36}$$

4. A four-dimensional finite element

From the general equations of the preceeding paragraph, any particular element with straight or curved boundaries may be developed to suit the boundaries of a gravitational field. In view of the numerical solution of nonlinear systems, we construct a finite element with simple geometry that forms topological simplexes in the space in which it is embedded [2]. For a k-dimensional space, such an element has $k + 1$ nodal points, thus, five in the considered case.

For this element, the local field $h^{\alpha\alpha}$ is approximated by functions linear in the coordinates x^i as following

$$h^{\alpha\alpha} = a^\alpha_0 + a^\alpha_i x^i\,, \tag{37}$$

where the a^α are unknown constants yet to be determined. On evaluating above expression at the five nodal points leads to the linear system

$$\boldsymbol{c}^\alpha \boldsymbol{a}^\alpha = \boldsymbol{G}^\alpha\,, \tag{38}$$

where the matrix $\boldsymbol{c}$ contains the nodal coordinates

$$\boldsymbol{c}^\alpha_\cdot = \begin{bmatrix} 1 & x^{11} \ldots x^{14} \\ 1 & x^{21} \quad\ x^{24} \\ 1 & x^{31} \quad\ x^{34} \\ 1 & x^{41} \quad\ x^{44} \\ 1 & x^{51} \ldots x^{54} \end{bmatrix}. \tag{39}$$

Here, the first superscript denotes the node number and the second one is the coordinate. Assuming det $(\boldsymbol{c}) \neq 0$, one may invert equation (38) to obtain

$$\boldsymbol{a}^\alpha = (\boldsymbol{c}^\alpha)^{-1}\, \boldsymbol{G}^\alpha \tag{40}$$

and substituting this equation back into equation (37) yields the desired local interpolation functions as

$$h^{\alpha\alpha} = \sum_{n=1}^{N} \Psi^\alpha_n \boldsymbol{G}^\alpha_n\,, \qquad N = 1, \ldots, 5\,, \tag{41}$$

where Ψ are linear functions in the coordinates x^i

$$\Psi^\alpha_n(\boldsymbol{x}) = A^\alpha_n + B^\alpha_{ni} x^i\,. \tag{42}$$

The constants A and B are elements of the inverted matrix $\boldsymbol{c}$ and may be expressed as

$$A^\alpha_n = \frac{1}{4!\, D}\, \varepsilon_{npqrs}\varepsilon_{ijkl} x^{pi} x^{qj} x^{rk} x^{sl}\,, \qquad B^\alpha_{ni} = \frac{-1}{4!\, D}\, \varepsilon_{npqrs}\varepsilon_{mjkl} \beta^{pm}_i \beta^{qj}_i \beta^{rk}_i \beta^{sl}_i\,, \tag{43}$$

$$n, p, q, r, s = 1, \ldots, 5\,, \quad i, j, k, l, m = 1, \ldots, 4\,,$$

where ε_{npqrs} and ε_{ijkl} are the five-dimensional and four-dimensional permutation symbols, respectively:

$$\beta_i^{rk} = \begin{cases} x^{rk} & \text{if} \quad i = k\,, \\ 1 & \text{if} \quad i \neq k \end{cases} \tag{44}$$

$$D = \det \mathbf{c} = 4!\, V\,, \tag{45}$$

where V is the volume of the simplex element. The functions Ψ have the following properties for every $x \in \mathcal{D}_e$:

$$\Psi_n^\alpha(x) \geqq 0\,, \qquad \sum_{n=1}^{5} \psi_n^\alpha(x) = 1\,,$$

and for any nodal point $x^m \in \mathcal{D}_e$

$$\psi_n^\alpha(x^m) = \delta_n^m\,. \tag{46}$$

The *element matrices* now take on the forms

$$E_{mn} = \int_{\mathcal{D}_e} [2\varepsilon^{\alpha\alpha}\delta_\alpha^i\delta_\alpha^j B_{mi}^\alpha B_{nj}^\alpha - \varepsilon^{\alpha\alpha}\varepsilon_{\beta\beta}\varepsilon_{\beta\beta}\delta_\alpha^i\delta_\alpha^j B_{mi}^\beta B_{nj}^\beta]\, \mathrm{d}\tau\,,$$

$$H_{mnk} = \int_{\mathcal{D}_e} (A_k^\beta + B_{kl}^\beta x^l)\,(2\varepsilon^{\alpha\alpha}\varepsilon_{\beta\beta}\delta_\alpha^i\delta_\alpha^j B_{mi}^\beta B_{nj}^\beta + \tfrac{1}{2}\,\varepsilon^{\alpha\alpha}\varepsilon_{\gamma\gamma}\varepsilon_{\gamma\gamma}\varepsilon_{\beta\beta}\delta_\alpha^i\delta_\alpha^j B_{mi}^\gamma B_{nj}^\gamma -$$

$$- 2\varepsilon_{\beta\beta}\varepsilon_{\beta\beta}\delta_\beta^i\delta_\beta^j B_{mi}^\beta B_{nj}^\beta - \varepsilon_{\alpha\alpha}\varepsilon_{\beta\beta}\delta_\alpha^i\delta_\alpha^j B_{mi}^\alpha B_{nj}^\alpha - \varepsilon_{\alpha\alpha}\varepsilon_{\alpha\alpha}\delta_\beta^i\delta_\beta^j B_{mi}^\alpha B_{nj}^\alpha)\, \mathrm{d}\tau\,,$$

$$Q_m = \int_{\mathcal{D}_e} [\varepsilon^{\alpha\alpha}\varepsilon^{\beta\beta}\delta_\beta^i B_{mi}^\alpha + \varepsilon^{\alpha\alpha}\varepsilon^{\beta\beta}(A_m^\alpha + B_{mi}^\alpha x^i)\, \Phi_{\alpha,\beta\beta}]\, \mathrm{d}\tau\,,$$

$$S_m = \int_{\mathcal{D}_e} (A_m^\alpha + B_{mi}^\alpha x^i)\, T_{\alpha\alpha}\, \mathrm{d}\tau \quad \text{and} \quad \delta_\alpha^i = \begin{cases} 1 & \text{if} \quad i = \alpha \\ 0 & \text{if} \quad i \neq \alpha\,. \end{cases} \tag{47}$$

5. A one-dimensional finite element

A time-independent, spheric-symmetrical gravitational field can be described by an one-dimensional finite element in radial direction r. In this case, equation (41) becomes

$$\begin{bmatrix} h^{11} \\ h^{22} \\ h^{33} \\ h^{44} \end{bmatrix} = \begin{bmatrix} \Psi_1^1 & \Psi_2^1 & & & & & & \\ & & \Psi_1^2 & \Psi_2^2 & & & & \\ & & & & \Psi_1^3 & \Psi_2^3 & & \\ & & & & & & \Psi_1^4 & \Psi_2^4 \end{bmatrix} \begin{bmatrix} G_1^1 \\ \vdots \\ \vdots \\ G_2^4 \end{bmatrix}. \tag{48}$$

The shape functions of equation (42) are expressable as

$$\Psi_1^\alpha = (r_2 - r)/L\,, \qquad \Psi_2^\alpha = (r - r_1)/L\,, \tag{49}$$

where $L = r_2 - r_1$ is the length of the element. That these functions are indeed suitable may easily be verified by the conditions (46). The substitution of the shape functions (49) into (20) yields the finite element matrices

$$E_{mn} = \int_{r_1}^{r_2} (-\Psi_{m,r}^1\Psi_{n,r}^1 + \Psi_{m,r}^2\Psi_{n,r}^2 + \Psi_{m,r}^3\Psi_{n,r}^3 + \Psi_{m,r}^4\Psi_{n,r}^4)\, \mathrm{d}r\,, \tag{50}$$

$$H_{mnk} = \tfrac{1}{2}\int_{r_1}^{r_2} [(-\Psi_k^1 + \Psi_k^2 + \Psi_k^3 - \Psi_k^4)\,(-\Psi_{m,r}^1\Psi_{n,r}^1 + \Psi_{m,r}^2\Psi_{n,r}^2 + \Psi_{m,r}^3\Psi_{n,r}^3 + \Psi_{m,r}^4\Psi_{n,r}^4)$$

$$- 4\Psi_{m,r}^1\Psi_{n,r}^1\Psi_k^1 + 4\Psi_{m,r}^2\Psi_{n,r}^2\Psi_k^2 + 4\Psi_{m,r}^3\Psi_{n,r}^3\Psi_k^3 - 4\Psi_{m,r}^4\Psi_{n,r}^4\Psi_k^4]\, \mathrm{d}r\,, \tag{51}$$

where a representative term has the integral

$$\int_{r_1}^{r_2} \Psi_{m,r}\Psi_{n,r}\Psi_k\, \mathrm{d}r = \tfrac{1}{2}\, L$$

and from equation (32)

$$Q_m = \int_{r_1}^{r_2} (\Psi_m^1\Phi_{,r}^1 + \Psi_m^2\Phi_{,r}^2 + \Psi_m^3\Phi_{,r}^3 - \Psi_m^4\Phi_{,r}^4)_{,r}\, \mathrm{d}r\,. \tag{52}$$

Choosing Newton's potential $\Phi = -M/r$ as boundary condition, the boundary reaction Q becomes

$$Q_m = -M \int_{r_1}^{r_2} [1/r^2\, (\Psi_m^1 + \Psi_m^2 + \Psi_m^3 - \Psi_m^4)]_{,r}\, \mathrm{d}r\,. \tag{53}$$

Based on this example, a computer program was written to solve the governing equations. The finite element results for the potentials $h^{\alpha\alpha}$ are compared with the Schwarzschild solution and are reported in table 1. The gravitational mass was chosen as $M = Gm/c^2 = 1$ cm and hence the distance r in table 1 measures also in cm. Starting with Newton's values as initial vector, the solution was found within 4 iterations.

The difference between the finite element results and the Schwarzschild solution arises from the weak field assumption of equation (10). However, with increasing distance from the gravitational mass; the difference of the two results becomes gradually smaller.

Table 1. Spherical gravitational field

Distance	SCHWARTZSCHILD		Finite element	
	g^{11}, g^{22}, g^{33}	g^{44}	g^{11}, g^{22}, g^{33}	g^{44}
40.000	−1.05000	.95000	−1.03446	.96554
50.000	−1.04000	.96000	−1.02940	.97060
75.000	−1.02667	.97333	−1.02233	.97767
125.000	−1.01600	.98400	−1.01588	.98412
200.000	−1.01000	.99000	−1.01101	.98899
300.000	−1.00667	.99333	−1.00673	.99327

References

1 SCHWARZSCHILD, K., Sitzungsbericht Preuss. Akad. Wiss. 1916, 424.
2 ODEN, J. T., Finite Elements of Nonlinear Continua, McGraw-Hill 1972.
3 GASS, N., Solution of the 3-dimensional scalar HELMHOLTZ equation by a finite element formulation, ZAMM 57, 95–99 (1977).
4 GASS, N., Solution of the SCHRÖDINGER wave equation in a space-time continuum by a finite element formulation, ZAMM 57, 285–291 (1977).
5 GASS, N., Some two-field variational principles for nonlinear deformation analysis of shells, ZAMM 55, 515–521 (1975).
6 WEBER, J., General Relativity and Gravitational Waves, Interscience Publishers, Inc., New York, 1961.
7 EINSTEIN, A.; LORENTZ, H. A.; WEYL, H.; MINKOWSKI, H., The Principle of Relativity, Dover Publication Inc. 1963.

Eingereicht: 8. 3. 1978, revidierte Fassung 15. 4. 1979

Anschrift: Dr. NORBERT GASS, 2960 Rue Longfellow, Sainte Foy, Québec G1W 1X7, Canada

ZAMM 60, 89–98 (1980)

W. SZABLEWSKI

Ähnlichkeitsgesetz des Spektrums von skalaren Schwankungen für den Trägheits-Konvektionsbereich $k \gg k_0$ in homogener isotroper Turbulenz

Auf der Grundlage der Formel von Corrsin für den Transfer skalarer Schwankungen, die äquivalent ist dem —5/3-Gesetz des skalaren Spektrums, wird für den Trägheits-Konvektionsbereich $k \gg k_0$ (k_0 Wellenzahl der großen Wirbel) ein Ähnlichkeitsgesetz des skalaren Spektrums hergeleitet, das die Modellkonstante Re_λ enthält und in der Asymptote in das —5/3-Gesetz von Obuchow-Corrsin übergeht. Für die empirischen Koeffizienten α bzw. α_θ des —5/3-Gesetzes des Spektrums der Turbulenzenergie bzw. des Spektrums der skalaren Schwankungen nehmen wir die Relation $\alpha_\theta = (2/3)\,\alpha$ an. Mit dem von Shih-Chun Lin, Shao-Chi Lin gemessenem eindimensionalen skalaren Spektrum bei $Re_\lambda = 150$ zeigt sich gute Übereinstimmung.

On the basis of Corrsin's formula for the transfer of scalar fluctuations — equivalent to the —5/3-power law of the scalar spectrum — a law of similarity is derived for the scalar spectrum in the inertial-convective region $k \gg k_0$ (k_0 wave number of the big eddies), which contains the Reynolds number Re_λ and has the asymptotic form of Obuchow-Corrsin's —5/3-power law. For the empirical coefficients α resp. α_θ of the —5/3-power law of the spectrum of the turbulent energy resp. of the scalar fluctuations we suppose the relation $\alpha_\theta = (2/3)\,\alpha$. Agreement with the measurements (Shih-Chun Lin and Shao-Chi Lin) of the one-dimensional scalar spectrum at $Re_\lambda = 150$.

На основании формулы Коррсина для переноса скалярных флуктуаций, эквивалентной —5/3-степенному закону скалярного спектра, выводится подобный закон скалярного спектра для инерционно-конвективной области $k \gg k_0$ (k_0-волновое число большого вихря), в который входит число Рейнольдса Re_λ и который в асимптотике переходит в —5/3-зтепенной закон Обухова-Коррсина. Для эмпирических коэффициентов α и α_θ —5/3-степенного закона спектра энергии турбулентности и спектра скалярных флуктуаций соответственно соотношение $\alpha_\theta = (2/3)\,\alpha$. Измеренный Ши-Хун Лином и Шао-Хи Лином одномерный скалярный спектр показал хорошее совпадение при $Re_\lambda = 150$.

Einleitung

Beim heutigen Stand der Theorie turbulenter Strömungen steht hinsichtlich der Struktur der Turbulenz das Modell homogener isotroper Turbulenz immer noch im Mittelpunkt der Forschung. Neben das zentrale Problem des Spektrums der Turbulenzenergie stellt sich hier das Problem des Spektrums von skalaren Schwankungen (Temperatur- und Konzentrationsschwankungen).

In einer vorangehenden Arbeit [1], auf die wir uns im folgenden mehrfach beziehen werden, war vom Verfasser auf der Grundlage der Formel von KOVASZNAY für den turbulenten Energiefluß, die äquivalent ist dem KOLMOGOROVschen —5/3-Gesetz des Spektrums der Turbulenzenergie, für den Trägheitsbereich $k \gg k_0$ (k_0 Wellenzahl der großen Wirbel) ein Ähnlichkeitsgesetz des Spektrums der Turbulenzenergie hergeleitet worden, das die Modellkonstante Re_λ enthält und in der Asymptote in das —5/3-Gesetz übergeht. Das entsprechende eindimensionale Spektrum stand in Übereinstimmung mit der neueren Messung von J. SCHEDVIN et al. [2].

Im vorliegenden Beitrag formulieren wir ein Ähnlichkeitsgesetz des Spektrums von skalaren Schwankungen im Trägheits-Konvektionsbereich $k \gg k_0$.

A. Gleichung des skalaren Spektrums

Frequenzanalyse von turbulenten Strömungen mit einem Temperaturgradienten oder einem Gradienten der Konzentration eines Stoffes — z. B. in der Atmosphäre und ozeanischen Gewässern; ferner in Prozessen der chemischen Verfahrenstechnik — ergibt ein kontinuierliches Spektrum der im quadratischen Mittel genommenen Temperatur — oder Konzentrationsschwankungen

$$\overline{T'^2} = \int_0^\infty \Theta_k \, \mathrm{d}k\,, \qquad k \text{ Wellenzahl;} \tag{1}$$

($\boldsymbol{k}$ Vektor im Wellenzahlenraum mit den Komponenten k_1, k_2, k_3) oder

$$\overline{c'^2} = \int_0^\infty \Gamma_k \, \mathrm{d}k\,.$$

Wir beziehen uns im folgenden auf skalare Schwankungen in abklingender homogener und isotroper Turbulenz (z. B. im turbulenten Nachlauf hinter einem beheizten Gitter).

a) Dreidimensionales Spektrum

Aus der Gleichung des Wärme- bzw. Stofftransports in inkompressibler Strömung

$$\frac{\partial T}{\partial t} = -\frac{\partial T u_k}{\partial x_k} + \frac{\lambda}{\varrho c_p} \nabla^2 T \tag{2}$$

bzw.

$$\frac{\partial c}{\partial t} = -\frac{\partial c u_k}{\partial x_k} + D\,\nabla^2 c$$

(λ Wärmeleitzahl, D Diffusionskoeffizient)

gewinnt man — vgl. z. B. I. O. Hinze [3] — über die aus (2) abzuleitende Gleichung für die Zweipunkt-Korrelation der Temperaturschwankungen in homogener isotroper Turbulenz

$$Q(\mathfrak{r}, t) = \overline{T'(\mathfrak{x})\, T'(\mathfrak{x} + \mathfrak{r})} \tag{3}$$

und Anwendung von Fouriertransformation die Gleichung für das Spektrum Θ_k

$$\frac{\partial \Theta_k}{\partial t} = -\frac{\partial F(k)}{\partial k} - 2\frac{\lambda}{\varrho c_p} k^2 \Theta_k\,, \tag{4}$$

integriert

$$\frac{\partial}{\partial t}\int\limits_0^k \Theta_k\,\mathrm{d}k = -F(k) - 2\frac{\lambda}{\varrho c_p}\int\limits_0^k k^2\Theta_k\,\mathrm{d}k\,. \tag{5}$$

Hier bezeichnet $F(k)$ symbolisch die Transformierte der nichtlinearen Glieder der Konvektion.

Das Problem der Konzentrationsschwankungen ist formal identisch mit dem der Temperaturschwankungen, so daß wir uns auf diese beschränken können.

Das zweite Glied auf der rechten Seite von (5) beschreibt die Dämpfung der turbulenten Temperaturschwankungen im Wellenzahlenbereich $< k$ durch die molekulare Wärmeleitung des Mediums; demnach ist das erste Glied der rechten Seite als (normierter) turbulenter Wärmefluß in den Wellenzahlenbereich $> k$ aufzufassen.

Da die Dämpfung der Temperaturschwankungen letztlich aus der molekularen Wärmeleitung resultiert,

$$\frac{\mathrm{d}}{\mathrm{d}t}\int\limits_0^\infty \Theta_k\,\mathrm{d}k\left(=\frac{\mathrm{d}\overline{T'^2}}{\mathrm{d}t}\right) = -2\frac{\lambda}{\varrho c_p}\int\limits_0^\infty k^2\Theta_k\,\mathrm{d}k \tag{6}$$

($-\mathrm{d}\overline{T'^2}/\mathrm{d}t = \varepsilon_\Theta$ sog. skalare Dissipation), haben wir für $F(k)$ die Randbedingung

$$\lim_{k\to\infty} F(k) = 0\,. \tag{7}$$

Die theoretische Beschreibung des skalaren Spektrums (Temperatur, Konzentration) stellt sich neben das Problem der theoretischen Beschreibung des Spektrums E_k der Turbulenzenergie E mit der Gleichung

$$\frac{\partial E_k}{\partial t} = -\frac{\partial S_k}{\partial k} - 2\nu k^2 E_k\,, \tag{8}$$

wo $S(k)$ symbolisch die Transformierte der nichtlinearen Glieder der Trägheitskräfte bezeichnet.

Die Problematik hinsichtlich der Gleichungen (4) und (8) liegt derzeit in der Aufgabe, den turbulenten Energiefluß $S(k)$ und den turbulenten Wärmefluß $F(k)$ in Relation zu den Variablen k, E_k und Θ_k zu setzen.

Hier liegen in bezug auf $F(k)$ bisher nur Hypothesen vor; vgl. z. B. I. O. Hinze [3]. Unter der Voraussetzung großer Pécletscher Zahl

$$Pe_{k_0} = \frac{\sqrt{E_{k_0}/k_0}}{\lambda/\varrho c_p} = Re_{k_0}\cdot Pr\,, \tag{9}$$

$$Re_{k_0} = \frac{\sqrt{E_{k_0}/k_0}}{\nu}\,, \qquad Pr = \frac{\varrho c_p \nu}{\lambda} \quad \text{(Prandtlzahl)}$$

des Spektrums, wo k_0 die Wellenzahl der großen Wirbel bezeichnet, läßt sich jedoch für einen Teilbereich des Spektrums eine definitive Aussage treffen. (Beim Problem der Konzentrationsschwankungen tritt an Stelle von Pr die Schmidtzahl $Sc = \nu/D$).

b) Eindimensionales Spektrum

Gemessen werden nur eindimensionale Spektren in Strömungsrichtung. Dreidimensionales Spektrum E_k und eindimensionales Spektrum

$$E_{11}(k_1)\left(\overline{u'^2} = \int\limits_0^\infty E_{11}(k_1)\,\mathrm{d}k_1\right)$$

sind in homogener isotroper Turbulenz durch

$$E_{k_1} = \frac{1}{2}k_1^2\frac{\partial^2 E_{11}(k_1)}{\partial k_1^2} - \frac{1}{2}k_1\frac{\partial E_{11}(k_1)}{\partial k_1} \tag{10}$$

verbunden.

Für das eindimensionale Spektrum

$$\Theta_1(k_1) \qquad \left(\overline{T'^2} = \int_0^\infty \Theta_1(k_1)\, \mathrm{d}k_1\right) \tag{11}$$

haben wir die Relation zum dreidimensionalen Spektrum

$$\Theta_1(k_1) = \int_{k_1}^{\infty} \frac{\Theta_k}{k}\, \mathrm{d}k \tag{12}$$

bzw.

$$\Theta_{k_1} = -\, k_1 \frac{\partial \Theta_1(k_1)}{\partial k_1}. \tag{13}$$

B. Trägheits- und Konvektionsunterbereich (inertial and convective subrange)

Bei großer PÉCLETscher Zahl

$$Pe_k = \frac{\sqrt{E_k/k}}{\lambda/\varrho c_p}$$

der Turbulenzelemente ist der Einfluß der molekularen Wärmeleitung als vernachlässigbar und bei abklingender Turbulenz das lokale Abklingen der Temperaturschwankungen im wesentlichen als durch den turbulenten Wärmefluß in den Wellenzahlenbereich $> k$ bedingt anzusehen. Wir können demnach bei großem Pe_{k_0} im Spektrum einen Konvektionsbereich (convective range) mit vernachlässigbarem Einfluß der Diffusion (molekulare Wärmeleitung) abgrenzen gegen einen Diffusionsbereich (diffusive range) mit merkbarem Einfluß der Diffusion. Mit wachsender PÉCLETscher Zahl Pe_{k_0} wird dabei der Diffusionsbereich des skalaren Spektrums, dessen Breite im Rahmen der Kontinuumsmechanik eine endliche Schranke gesetzt ist, immer mehr eingeengt.

Bei hinreichend großem Pe_{k_0} wird es dann weiter — vgl. die analoge Betrachtung in [1] — im Konvektionsbereich einen an den Diffusionsbereich angrenzenden Unterbereich (convective subrange) geben, für welchen

α) Unabhängigkeit von den äußeren Bedingungen bzw. von den großen Wirbeln k_0 angenommen werden kann;

β) der turbulente Wärmefluß hier

$$F(k) = -\frac{\partial}{\partial t}\int_0^k \Theta_k\, \mathrm{d}k \approx -\frac{\mathrm{d}}{\mathrm{d}t}\int_0^\infty \Theta_k\, \mathrm{d}k = \varepsilon_\Theta \tag{14}$$

gesetzt werden kann.

Handelt es sich um einen stationären Zustand, so gilt

$$\varepsilon_\Theta \approx F(k) + 2\frac{\lambda}{\varrho c_p}\int_0^k k^2 \Theta_k\, \mathrm{d}k \approx F(k)\,. \tag{15}$$

Entsprechend ergibt sich bei hinreichend großer REYNOLDSscher Zahl Re_{k_0}, was wir voraussetzen wollen, für den Trägheitsunterbereich

$$S(k) = -\frac{\partial}{\partial t}\int_0^k E_k\, \mathrm{d}k \approx -\frac{\mathrm{d}}{\mathrm{d}t}\int_0^\infty E_k\, \mathrm{d}k = \varepsilon; \tag{16}$$

bei stationärem Zustand

$$\varepsilon \approx S(k) + 2\nu \int_0^k k^2 E_k\, \mathrm{d}k \approx S(k)\,. \tag{17}$$

Es folgt dann für den Trägheitsunterbereich dimensionsanalytisch nach KOLMOGOROV [4] das Spektrum der Turbulenzenergie

$$E_k \approx \alpha \varepsilon^{2/3} k^{-5/3} \tag{18}$$

bzw.

$$E_k = \alpha \varepsilon^{2/3} k^{-5/3} \quad \text{für} \quad k \to \infty\,. \tag{19}$$

Im Konvektionsunterbereich ist $\Theta_k[T^2\,\mathrm{m}]$ zunächst einmal als durch die Wellenzahl $k[1/\mathrm{m}]$ und die skalare Dissipation $\varepsilon_\Theta[T^2/\mathrm{sec}]$ determiniert anzusehen. Dimensionsanalyse zeigt, daß noch ein charakteristischer Parameter, der t enthält, heranzuziehen ist, als welcher sich hier ε bietet. Man erhält dann das skalare Spektrum

$$\Theta_k \approx \alpha_\Theta \varepsilon_\Theta \varepsilon^{-1/3} k^{-5/3} \tag{20}$$

(α_Θ empirische dimensionslose Konstante) bzw.

$$\Theta_k = \alpha_\Theta \varepsilon_\Theta \varepsilon^{-1/3} k^{-5/3} \quad \text{für} \quad k \to \infty\,. \tag{21}$$

Dieses Spektralgesetz wurde unabhängig voneinander von A. M. Obuchow [5] und S. Corrsin [6] hergeleitet und ist durch Messungen bestätigt worden (vgl. auch Abb. 4).

Für die eindimensionalen Spektren erhält man dann nach (10) und (13)

$$E_{11}(k_1) = \alpha_1 \varepsilon^{2/3} k_1^{-5/3} \quad \text{mit } \alpha_1 = \frac{18}{55}\alpha\,, \tag{22}$$

$$\Theta_1(k_1) = \alpha_{\Theta_1} \varepsilon_\Theta \varepsilon^{-1/3} k_1^{-5/3} \quad \text{mit } \alpha_{\Theta_1} = \frac{3}{5}\alpha_\Theta\,. \tag{23}$$

Als wichtige Folgerung ergab sich — vgl. [1] — aus (18) mit $S \approx \varepsilon$ der Ansatz von Kovasznay [7]

$$S(k) \approx \alpha^{-3/2} E_k^{3/2} k^{5/2} \tag{24}$$

im Trägheitsunterbereich.

Entsprechend folgt aus (21) mit $F \approx \varepsilon_\Theta$ und $\varepsilon \approx \alpha^{-3/2} E_k^{3/2} k^{5/2}$ nach (18) ein Ansatz von S. Corrsin [17], der damit gegenüber anderen Formulierungen eine Bestätigung findet,

$$F(k) \approx \alpha_\Theta^{-1} \alpha^{-1/2} \Theta_k E_k^{1/2} k^{5/2} \tag{25}$$

im Konvektionsunterbereich.

C. Universeller Bereich

Beim Übergang vom Trägheitsunterbereich in den Zähigkeitsbereich tritt als weitere determinierende physikalische Einflußgröße die kinematische Zähigkeit ν und damit neben $1/k$ die weitere determinierende Länge $(\nu^3/\varepsilon)^{1/4}$ auf. Dimensionsanalysis führt dann bei Fortsetzung von (18) in den Zähigkeitsbereich und damit für den sogenannten universellen Bereich nach Kolmogorov bei Einführung der dimensionslosen Größen

$$\tilde{k} = k(\nu^3/\varepsilon)^{1/4}\,, \qquad \tilde{E}_{\tilde{k}} = E_k(\nu^5 \cdot \varepsilon)^{-1/4} \tag{26}$$

auf die Form

$$\tilde{E}_{\tilde{k}} = \alpha \tilde{k}^{-5/3} f(\tilde{k}) \tag{27}$$

mit der dimensionsanalytisch unbestimmt bleibenden Funktion $f(\tilde{k})$. Entsprechend für den Fluß kinetischer Energie im universellen Bereich

$$\tilde{S}(\tilde{k})\ (\equiv S/\varepsilon) = \alpha^{-3/2} \tilde{E}_{\tilde{k}}^{3/2} \tilde{k}^{5/2} f_1(\tilde{k})\,. \tag{28}$$

Die Gleichung (8) des Spektrums der Turbulenzenergie nimmt dann — vgl. [1] — im universellen Bereich die Form

$$a\left(\tilde{k}\,\frac{d\tilde{E}_{\tilde{k}}}{d\tilde{k}} - \tilde{E}_{\tilde{k}}\right) = -\frac{d\tilde{S}}{d\tilde{k}} - 2\tilde{k}^2 \tilde{E}_{\tilde{k}} \tag{29}$$

an mit der Dissipation

$$\varepsilon = \frac{1}{4}\,\frac{\nu}{a^2}\,\frac{1}{t^2} \tag{30}$$

und — bei Nullsetzen der Integrationskonstante — dem Abfall der Turbulenzenergie

$$E = \frac{1}{4}\,\frac{\nu}{a^2}\,\frac{1}{t}\,. \tag{31}$$

Mit der turbulenten Reynoldszahl

$$Re_\lambda = \frac{\sqrt{\overline{u'^2}}\,\lambda}{\nu}$$

($\lambda = \sqrt{15\nu\overline{u'^2}/\varepsilon}$ Taylorsche Mikrolänge), die eine Modellkonstante darstellt, steht a in der Beziehung

$$a = \sqrt{5/3}\ Re_\lambda^{-1}\,. \tag{32}$$

Nimmt man an, daß für $\tilde{k} \to \infty$ mit $Re_k \to 0$ in der lokalen Bilanz der Turbulenzenergie der durch den Wirbelumsatz bedingte Beitrag $d\tilde{S}/d\tilde{k}$ gegenüber der viskosen Dissipation als vernachlässigbar anzusehen ist, so folgt aus (29) für die Asymptote $\tilde{k} \to \infty$

$$\tilde{E}_{\tilde{k}} = \beta \tilde{k}\, e^{-\tilde{k}^2/a} \tag{33}$$

(β dimensionslose empirische Konstante) bzw.

$$E_k = \beta \nu^2 k\, e^{-2\nu k^2 t}\,. \tag{34}$$

Entsprechend erhalten wir bei Fortsetzung von (20) in den Zähigkeits- und Diffusionsbereich mit

$$\tilde{k}_\Theta \left(\equiv k((\lambda/\varrho c_p)^3/\varepsilon)^{1/4}\right) = \tilde{k} Pr^{-3/4} \tag{35}$$

die Form

$$\Theta_k = \alpha_\Theta \varepsilon^{-1/3} \varepsilon_\Theta k^{-5/3} g(\tilde{k}) \,, \tag{36}$$

wobei die Abhängigkeit von der PRANDTLzahl Pr in $g(\tilde{k})$ eingeschlossen ist. Führen wir die weitere dimensionslose Größe

$$\tilde{\Theta}_{\tilde{k}} = \Theta_k \varepsilon^{3/4} \varepsilon_\Theta^{-1} \nu^{-5/4} \tag{37}$$

ein, so gilt

$$\tilde{\Theta}_{\tilde{k}} = \alpha_\Theta \tilde{k}^{-5/3} g(k) \tag{38}$$

im universellen Bereich $k \gg k_0$. Entsprechend für den Fluß von Wärmeenergie im universellen Bereich

$$\tilde{F}(\tilde{k}) \left(\equiv F(k)/\varepsilon_\Theta\right) = \alpha^{-1/2} \alpha_\Theta^{-1} E_{\tilde{k}}^{1/2} \tilde{\Theta}_{\tilde{k}} k^{5/2} g_1(\tilde{k}) \,. \tag{39}$$

Für $\tilde{\Theta}_{\tilde{k}}$ nimmt Gleichung (4) unter Beachtung von

$$\partial\Theta_k/\partial t|_k = \nu^{5/4} \left\{ \frac{\mathrm{d}(\varepsilon_\Theta/\varepsilon^{3/4})}{\mathrm{d}t} \tilde{\Theta}_{\tilde{k}} - \frac{1}{4} \left(\frac{\varepsilon_\Theta}{\varepsilon^{3/4}}\right) \frac{1}{\varepsilon} \frac{\mathrm{d}\varepsilon}{\mathrm{d}t} \tilde{k} \frac{\mathrm{d}\tilde{\Theta}_{\tilde{k}}}{\mathrm{d}\tilde{k}} \right\}$$

dann die Form an

$$\frac{\nu^{1/2}\varepsilon^{1/4}}{\varepsilon_\Theta} \left(\frac{\mathrm{d}(\varepsilon_\Theta/\varepsilon^{3/4})}{\mathrm{d}t} \tilde{\Theta}_{\tilde{k}} - \frac{1}{4} \frac{\varepsilon_\Theta}{\varepsilon^{3/4}} \frac{1}{\varepsilon} \frac{\mathrm{d}\varepsilon}{\mathrm{d}t} \tilde{k} \frac{\mathrm{d}\tilde{\Theta}_{\tilde{k}}}{\mathrm{d}\tilde{k}} \right) = - \frac{\mathrm{d}\tilde{F}(\tilde{k})}{\mathrm{d}\tilde{k}} - \frac{2}{Pr} \tilde{k}^2 \tilde{\Theta}_{\tilde{k}} \,.$$

Berücksichtigt man die Relation (30), so erhält man

$$\left(\left(\frac{\nu}{\varepsilon}\right)^{1/2} \frac{1}{\varepsilon_\Theta} \frac{\mathrm{d}\varepsilon_\Theta}{\mathrm{d}t} + 3a \right) \tilde{\Theta}_{\tilde{k}} + a\tilde{k} \frac{\mathrm{d}\tilde{\Theta}_{\tilde{k}}}{\mathrm{d}\tilde{k}} = - \frac{\mathrm{d}\tilde{F}(\tilde{k})}{\mathrm{d}\tilde{k}} - \frac{2}{Pr} \tilde{k}^2 \tilde{\Theta}_{\tilde{k}} \,. \tag{40}$$

Es folgt demnach für den universellen Bereich mit $(\nu/\varepsilon)^{1/2} = 2at$

$$2at \frac{1}{\varepsilon_\Theta} \frac{\mathrm{d}\varepsilon_\Theta}{\mathrm{d}t} = \text{const.}\,,$$

also ein Potenzgesetz

$$\varepsilon_\Theta \sim t^{-(n+1)} \,,$$

bzw. bei Nullsetzen der Integrationskonstante (41)

$$\overline{\Theta'^2} \sim t^{-n} \,.$$

Damit erhalten wir schließlich für den universellen Bereich die Gleichung

$$a \left\{ (-2n + 1)\, \tilde{\Theta}_{\tilde{k}} + \tilde{k} \frac{\mathrm{d}\tilde{\Theta}_{\tilde{k}}}{\mathrm{d}\tilde{k}} \right\} = - \frac{\mathrm{d}\tilde{F}(\tilde{k})}{\mathrm{d}\tilde{k}} - \frac{2}{Pr} \tilde{k}^2 \tilde{\Theta}_{\tilde{k}} \,. \tag{42}$$

Nehmen wir wieder an, daß für $\tilde{k} \to \infty$ mit Pe_k $(= Re_k \cdot Pr) \to 0$ in der lokalen Bilanz der durch den Wirbelumsatz bedingte Beitrag zur Wärmeenergie vernachlässigbar ist gegen die skalare Dissipation, so folgt aus (42) die asymptotische Formel

$$\tilde{\Theta}_{\tilde{k}} = \beta_\Theta \tilde{k}^{2n-1} \, \mathrm{e}^{-(1/Pr)\cdot \tilde{k}^2/a} \tag{43}$$

(β_Θ dimensionslose empirische Konstante) bzw.

$$\Theta_k = \beta_\Theta (\varepsilon_\Theta t^{n+1}) \, (2a)^{n+1} \, \nu^n k^{2n-1} \, \mathrm{e}^{(-2\nu/Pr)\cdot k^2 t} \,. \tag{44}$$

Wäre

$$\frac{-\mathrm{d}\tilde{F}(\tilde{k})/\mathrm{d}\tilde{k}}{(2/Pr)\, \tilde{k}^2 \tilde{\Theta}_{\tilde{k}}} \to \gamma_\Theta \qquad \text{für} \quad \tilde{k} \to \infty \,, \tag{45}$$

so müßte $0 \leqq \gamma_\Theta < 1$ sein und man erhielte asymptotisch

$$\tilde{\Theta}_{\tilde{k}} \sim \tilde{k}^{2n-1} \, \mathrm{e}^{-(1-\gamma_\Theta) \frac{1}{Pr} \tilde{k}^2/a} \tag{46}$$

Hinsichtlich des Exponenten n des Potenzgesetzes für den Abfall des skalaren Feldes liegt bisher eine theoretische Aussage von I. O. HINZE [3] vor: Unter der Voraussetzung, daß $F(k)$ sich *analytisch* bei $k = 0$ verhält bzw. nach Potenzen von k^2 entwickelt werden kann, läßt sich die Invariante

$$\overline{T'^2}\, \lambda_\Theta^3 = \text{const.} \qquad \left(\lambda_\Theta = \sqrt{\frac{12}{n} \, (\lambda/\varrho c_p)\, t} \right), \tag{47}$$

herleiten, die zusammen mit der Gleichung

$$\frac{d\overline{T'^2}}{dt} = -12 \frac{\lambda}{\varrho c_p} \frac{\overline{T'^2}}{\lambda_\Theta^2} \tag{48}$$

(Analogon zur Kármán-Howarthschen Gleichung der Turbulenzenergie) den Exponenten zu

$$n = 3/2 \tag{49}$$

bestimmt.

Vorliegende experimentelle Ergebnisse hinter beheizten Gittern zeigen unterschiedliche Werte für $\overline{T'^2} \sim x^{-n}$: R. R. Mills et al. [8] bzw. T. T. Yeh, C. W. Van Atta [9] erhielten $n \approx 0{,}87$ bzw. 1,29, wobei letztere Messung nahezu gleichen zeitlichen Abfall für die Geschwindigkeitsschwankungen ergab. Shih-Chun Lin, Shao-Chi Lin [10] schließlich erhielten bei relativ starker Intensität der Geschwindigkeitsschwankungen sogar Werte $n = 2{,}5$ bis $n = 3{,}1$. Messungen des Abfalls einer Salzkonzentration in einer Gitterströmung wurden von C. H. Gibson, W. H. Schwarz [11] durchgeführt, denen zufolge $n = 1{,}5$ ist.

Beim vorliegenden Modell handelt es sich um ein skalares Feld T ($T = \overline{T} + T'$) in einer abklingenden homogenen isotropen Strömung bei relativ großer Reynoldszahl Re_λ der Turbulenz (Frühstadium). Der Abbau der Temperatur — wie der Geschwindigkeitsschwankungen beim Fortschreiten der Strömung erfolgt *hier* wesentlich durch den *Kaskadenprozeß* (*Zerfall der Wirbel*) und anschließende skalare bzw. viskose Dissipation. Wir nehmen daher für das Frühstadium gleiches zeitliches Abklingen der skalaren wie der Geschwindigkeitsschwankungen an; d. h. wir setzen im folgenden nach (31)

$$n = 1\,. \tag{50}$$

D. Trägheits-Konvektionsbereich $k \gg k_0$

Im Konvektionsbereich $k \gg k_0$ ergibt Dimensionsanalysis auch unabhängig von der Existenz eines Konvektionsunterbereichs bzw. der Forderung $F(k) \approx \varepsilon_\Theta$

$$F(k) = \alpha^{-1/2} \alpha_\Theta^{-1} E_k^{1/2} \Theta_k k^{5/2}\,, \tag{51}$$

denn Θ_k und $F(k)$ bedingen offenbar einander, und im Konvektionsbereich $k \gg k_0$ steht nur die Länge $1/k$ zur Verfügung; unter der notwendigen Heranziehung von E_k folgt dann (51). Entsprechend hatte sich im Trägheitsbereich $k \gg k_0$ — vgl. [1] — ergeben

$$S(k) = \alpha^{-3/2} E_k^{3/2} k^{5/2}\,. \tag{52}$$

Für $k \to \infty$ erhält man dann wieder mit $S(k) \to \varepsilon$, $F(k) \to \varepsilon_\Theta$ die Spektralgesetze (19) und (21).

Setzt man (51) in (4) ein, so wird das Spektrum Θ_k im genannten Bereich durch die Gleichung

$$\frac{\partial \Theta_k}{\partial t} + \alpha^{-1/2} \alpha_\Theta^{-1} k^{5/2} E_k^{1/2} \frac{\partial \Theta_k}{\partial k} = \frac{\alpha^{-1/2}}{2} \alpha_\Theta^{-1} \left(k^{5/2} E_k^{-1/2} \frac{\partial E_k}{\partial k} + 5 k^{3/2} E_k^{1/2} \right) \Theta_k \tag{53}$$

beschrieben.

Aus Experimenten abgeleitete Werte zeigen eine starke Streuung der Koeffizienten α_Θ bzw. α_{Θ_1}: Messungen im Nachlauf einer Gitterströmung ergaben nach Gibson und Schwarz [11] $\alpha_{\Theta_1} = 0{,}35$ ($\alpha_1 = 0{,}44$), nach Lin und Lin [10] $\alpha_{\Theta_1} = 0{,}60$; Messungen in der Atmosphäre nach Gibson et al. [12] $\alpha_{\Theta_1} = 1{,}15$ ($\alpha_1 = 0{,}69$), nach Boston und Burling [13] $\alpha_{\Theta_1} = 0{,}81$ ($\alpha_1 = 0{,}51$), nach Williams und Paulson [14] $\alpha_{\Theta_1} = 0{,}50$ ($\alpha_1 = 0{,}54$); Messungen im Ozean nach Grant et al. [15] $\alpha_{\Theta_1} = 0{,}31$ ($\alpha_1 = 0{,}33$).

Im folgenden nehmen wir

$$\alpha_\Theta = (2/3)\,\alpha \tag{54}$$

an; d. h. wir nehmen im Wellenzahlenraum k gleichen Transport für die Skalare $\overline{\Theta'^2}$ und $\overline{u'^2}$ an.

Für das eindimensionale Spektrum nach (22, 23) gilt mithin

$$\alpha_{\Theta_1} = \frac{11}{9} \alpha_1 \quad \text{und} \quad \alpha_{\Theta_1} = \frac{2}{5} \alpha\,. \tag{55}$$

Setzen wir — wie in [1] — $\alpha = 1{,}6$, so erhalten wir

$$\alpha_{\Theta_1} = 0{,}64\,, \tag{56}$$

welchen Wert wir den nachfolgenden numerischen Ausführungen im konkreten Fall zugrundelegen.

1. Dreidimensionales Spektrum

Die Lösung des charakteristischen Systems der Gleichung (8) mit (24) ergab für den Trägheitsbereich $k \gg k_0$ — vgl. [1] — in der Variablen

$$\psi = \frac{1}{\alpha} \tilde{k}^{5/3} \tilde{E}_{\tilde{k}}\,, \qquad 0 \leqq \psi < 1$$

das Ähnlichkeitsgesetz $E_*(k_*)$ in der Parameterdarstellung

$$E_* \; (\equiv \widetilde{E}_{\widetilde{k}} M^{5/2}/\alpha) = \psi^{3/8}(1-\psi^{3/4})^{5/2} \,, \qquad k_* \; (\equiv \widetilde{k} M^{-3/2}) = \frac{\psi^{3/8}}{(1-\psi^{3/4})^{3/2}} \,, \tag{57}$$

mit

$$M = 2\alpha \sqrt{5/3}\; Re_\lambda^{-1} \equiv 2\alpha a$$

oder auch mit

$$x = \psi^{3/4} \,, \qquad 0 \leqq x < 1 \qquad E_* = \sqrt{x}\,(1-x)^{5/2} \,, \qquad k_* = \frac{\sqrt{x}}{(1-x)^{3/2}} \,. \tag{58}$$

Von diesem Ergebnis ausgehend legen wir die Ermittlung des skalaren Spektrums im Trägheits-Konvektionsbereich $k \gg k_0$ die Gleichung (42) des universellen Bereiches zugrunde, die hier

$$a \left\{ (1-2n)\, \widetilde{\Theta}_{\widetilde{k}} + \widetilde{k} \frac{\mathrm{d}\widetilde{\Theta}_{\widetilde{k}}}{\mathrm{d}\widetilde{k}} \right\} = - \frac{\mathrm{d}\widetilde{F}(\widetilde{k})}{\mathrm{d}\widetilde{k}} \tag{59}$$

lautet, wobei nach (39)

$$\widetilde{F}(\widetilde{k}) = \alpha^{-1/2} \alpha_\Theta^{-1} \widetilde{k}^{5/2} \widetilde{E}_{\widetilde{k}}^{1/2} \widetilde{\Theta}_{\widetilde{k}} \,.$$

Wir erhalten dann zunächst

$$\frac{\mathrm{d}\Theta_{\widetilde{k}}}{\widetilde{\Theta}_{\widetilde{k}}} = \frac{(2n-1)\,a - \alpha^{-1/2}\alpha_\Theta^{-1}(\mathrm{d}/\mathrm{d}\widetilde{k})\,(\widetilde{k}^{5/2}\widetilde{E}_{\widetilde{k}}^{1/2})}{a\widetilde{k} + \alpha^{-1/2}\alpha_\Theta^{-1}\widetilde{k}^{5/2}\widetilde{E}_{\widetilde{k}}^{1/2}}\,\mathrm{d}\widetilde{k}$$

bzw.

$$\ln \widetilde{\Theta}_{\widetilde{k}} = -\ln\left(a\widetilde{k} + \alpha^{-1/2}\alpha_\Theta^{-1}\widetilde{k}^{5/2}E_{\widetilde{k}}^{1/2}\right) + 2na \int \frac{\mathrm{d}\widetilde{k}}{a\widetilde{k} + \alpha^{-1/2}\alpha_\Theta^{-1}\widetilde{k}^{5/2}\widetilde{E}_{\widetilde{k}}^{1/2}} \,. \tag{60}$$

Führen wir die Variablen des Ähnlichkeitsgesetzes (58) ein,

$$a\widetilde{k} = aM^{3/2}k_* = aM^{3/2}\frac{\sqrt{x}}{(1-x)^{3/2}} \,,$$

$$\widetilde{k}^{5/2}\widetilde{E}_{\widetilde{k}}^{1/2} = \sqrt{\alpha}\,\sqrt{\psi}\; M^{5/2}k_*^{5/3} = \sqrt{\alpha}\; M^{5/2}\frac{x^{3/2}}{(1-x)^{5/2}} \,, \tag{61}$$

$$\mathrm{d}\widetilde{k} = M^{3/2}\frac{\mathrm{d}k_*}{\mathrm{d}x}\,\mathrm{d}x = \frac{M^{3/2}}{2}\,\frac{1+2x}{\sqrt{x}\,(1-x)^{5/2}}\,\mathrm{d}x \,,$$

so erhalten wir

$$\ln \widetilde{\Theta}_{\widetilde{k}} = \ln \frac{(1-x)^{5/2}}{\sqrt{x}\,(1+Ax)} + n \int \frac{1+2x}{Ax^2+x}\,\mathrm{d}x \tag{62}$$

mit $A = 2\dfrac{\alpha}{\alpha_\Theta} - 1$. Elementare Integration ergibt

$$\int \frac{1+2x}{Ax^2+x}\,\mathrm{d}x = \ln\left(\frac{x}{1+Ax}\right) + \frac{2}{A}\ln(1+Ax) \,,$$

und damit dann

$$\widetilde{\Theta}_{\widetilde{k}} = \mathrm{const}\,\frac{(1-x)^{5/2}}{\sqrt{x}\,(1+Ax)}\,x^n(1+Ax)^{\left(\frac{2}{A}-1\right)n} \,.$$

Zu fordern ist, daß für $x \to 1$ (d. h. $\widetilde{k} \to \infty$) nach (38)

$$\widetilde{\Theta}_{\widetilde{k}} = \alpha_\Theta \widetilde{k}^{-5/3} = \alpha_\Theta M^{-5/2}(1-x)^{5/2} \tag{63}$$

gilt. Erfüllung dieser Randbedingung ergibt dann schließlich für

$$\Theta_*(k_*) = \widetilde{\Theta}_{\widetilde{k}} M^{5/2}/\alpha_\Theta \tag{64}$$

das *Ähnlichkeitsgesetz des skalaren Spektrums im Trägheits-Konvektionsbereich* $k \gg k_0$ in der Parameterdarstellung

$$\begin{aligned} \Theta_*(k_*) &= (1-x)^{5/2}\, x^{n-1/2} \left(\frac{1+A}{1+Ax}\right)^{1-n\left(\frac{2}{A}-1\right)} \\ k_* &= \frac{\sqrt{x}}{(1-x)^{3/2}}\,; \qquad 0 \leqq x < 1\,. \end{aligned} \tag{65}$$

Bei $\alpha_\Theta = \dfrac{2}{3}\,\alpha$ nach (54) bzw. mit $A = 2$ gilt demnach

$$\Theta_*(k_*) = (1-x)^{5/2}\, x^{n-1/2} \cdot \frac{3}{1+2x}\,. \tag{66}$$

In Abb. 1 haben wir die erhaltenen spektralen Verteilungen $E_*(k_*)$ nach (58) und $\Theta_*(k_*)$ nach (66) für $n = 1$ (vgl. (50)), $n = 3/2$ (vgl. (49)) und $n = 5/2$ (Messung LIN/LIN) aufgetragen. Auffallend ist die größere Konsistenz des $-5/3$-Gesetzes des skalaren Spektrums für $n = 1$.

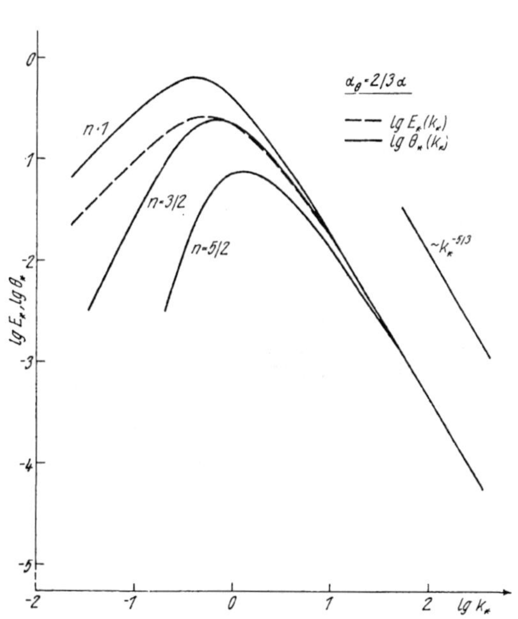

Abb. 1. Spektren der Turbulenzenergie und der skalaren Schwankungen im Trägheits-Konvektionsbereich $k \gg k_0$

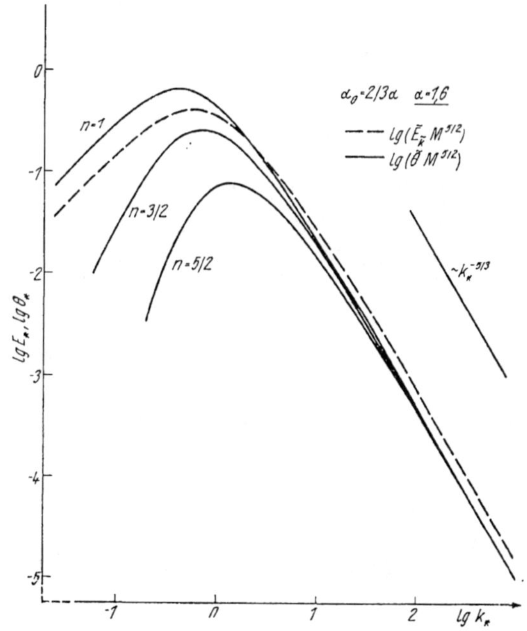

Abb. 2. Verlauf von $\widetilde{E}_{\widetilde{k}} M^{5/2}$ und $\widetilde{\Theta}_{\widetilde{k}} M^{5/2}$ im Trägheits-Konvektionsbereich $k \gg k_0$ (lies lg $\widetilde{E}_{\widetilde{k}} M^{5/2}$, lg $\widetilde{\Theta}_{\widetilde{k}} M^{5/2}$)

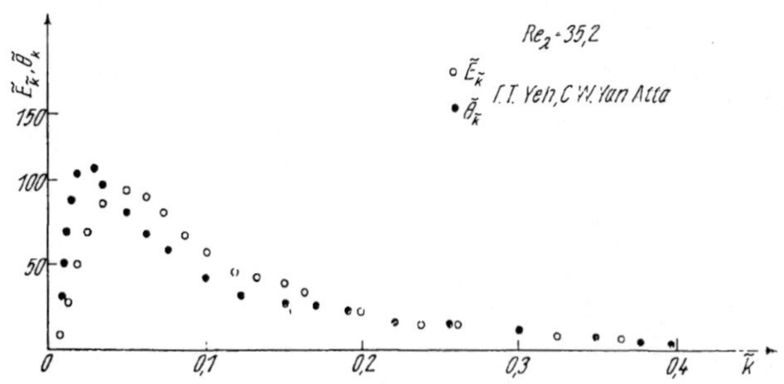

Abb. 3. Dreidimensionale Spektren der Turbulenzenergie und der skalaren Schwankungen bei $Re_\lambda = 35{,}2$

Wie in einem konkreten Fall die Verteilungen $\widetilde{E}_{\widetilde{k}}$ und $\widetilde{\Theta}_{\widetilde{k}}$ zueinander liegen, ist aus Abb. 2 zu ersehen, in der für $\alpha = 1{,}6$ (wie in [1]) und $\alpha_\Theta = (2/3)\,\alpha$ die Verteilungen $\widetilde{E}_k M^{5/2} = \alpha E_*(k_*)$ und $\widetilde{\Theta}_{\widetilde{k}} M^{5/2} = (2/3)\,\alpha \Theta_*(k_*)$ nach (66) dargestellt sind. Demnach verläuft bei $n = 1$ $\widetilde{\Theta}_{\widetilde{k}}$ für größere k zunächst unterhalb $\widetilde{E}_{\widetilde{k}}$ (entsprechend den $-5/3$-Gesetzen $\widetilde{E}_{\widetilde{k}} = \alpha k^{-5/3}$ und $\widetilde{\Theta}_{\widetilde{k}} = \alpha_\Theta k^{-5/3}$ mit $\alpha_\Theta < \alpha$), um dann jedoch für kleinere k oberhalb von $\widetilde{E}_{\widetilde{k}}$ zu liegen. Dieser Sachverhalt wird durch die von YEH und VAN ATTA [6] aus ihren Messungen eindimensionaler Spektren gewonnenen dreidimensionalen Spektren qualitativ bestätigt (Abb. 3). Für einen quantitativen Vergleich von Theorie und Experiment fehlen hier allerdings die Voraussetzungen: Die vorliegende Theorie setzt relativ große REYNOLDSsche Zahl Re_λ bzw. die Existenz des $-5/3$-Gesetzes der Turbulenzenergie voraus. Demgegenüber wurde die obige Messung bei der relativ kleinen REYNOLDSzahl $Re_\lambda = 35{,}2$ durchgeführt, und ein $-5/3$-Gesetz der Turbulenzenergie ist dann auch, wie in [9] angegeben wird, nicht feststellbar. Wir vermerken in dieser Beziehung noch, daß bei der relativ großen REYNOLDSzahl $Re_\lambda = 280$ (Messung einer Gitterströmung nach SCHEDVIN [2]) der quantitative Vergleich mit dem gemessenen eindimensionalen Spektrum der Turbulenzenergie im Trägheitsbereich $k \gg k_0$ völlige Übereinstimmung ergab (vgl. [1]).

2. Eindimensionales Spektrum

Das eindimensionale Spektrum $\tilde{\Theta}_1(\tilde{k}_1)$ steht nach (12) zum dreidimensionalen Spektrum $\tilde{\Theta}(\tilde{k})$ in der Beziehung

$$\tilde{\Theta}_1(\tilde{k}_1) = \int\limits_{\tilde{k}_1}^{\infty} \frac{\Theta_{\tilde{k}}}{\tilde{k}} \,\mathrm{d}\tilde{k} \quad \text{bzw.} \quad \Theta_{*1}(k_{*1}) = \int\limits_{k_{*1}}^{\infty} \frac{\Theta_*(k_*)}{k_*} \,\mathrm{d}k_* \quad \text{mit} \quad \frac{dk_*}{k_*} = \frac{1}{2}\,\frac{1}{x}\,\frac{1+2x}{1-x}\,\mathrm{d}x\,. \tag{67}$$

Damit gilt also nach (65)

$$\Theta_{*1} = \frac{1}{2}\int\limits_{x_1}^{1} (1-x)^{3/2}\, x^{n-3/2}(1+2x)\left(\frac{1+A}{1+Ax}\right)^{1-n\left(\frac{2}{A}-1\right)} \mathrm{d}x\,. \tag{68}$$

Bei $\alpha_\Theta = \frac{2}{3}\alpha$ bzw. $A = 2$ wird demnach

$$\Theta_{*1} = \frac{3}{2}\int\limits_{x_1}^{1} (1-x)^{3/2}\, x^{n-3/2}\,\mathrm{d}x \quad \text{mit} \quad k_{*1} = \frac{\sqrt{x_1}}{(1-x_1)^{3/2}}\,. \tag{69}$$

Wir erhalten dann nach (69) für $n = 1$, $n = 3/2$ und $n = 5/2$ die dem Ähnlichkeitsgesetz entsprechenden eindimensionalen Spektren

$$\begin{aligned} n = 1:&\quad \Theta_{*1} = \frac{3}{2}\left\{\sqrt{x_1(1-x_1)}\left[-\frac{5}{4}+\frac{1}{2}x_1\right]+\frac{3}{4}\operatorname{arctg}\sqrt{\frac{1-x_1}{x_1}}\right\},\\ n = \frac{3}{2}:&\quad \Theta_{*1} = \frac{3}{5}(1-x_1)^{5/2}\,,\\ n = \frac{5}{2}:&\quad \Theta_{*1} = \frac{3}{5}(1-x_1)^{5/2}\left[1-\frac{5}{7}(1-x_1)\right], \end{aligned} \tag{70}$$

die wir in Abb. 4 aufgetragen haben.

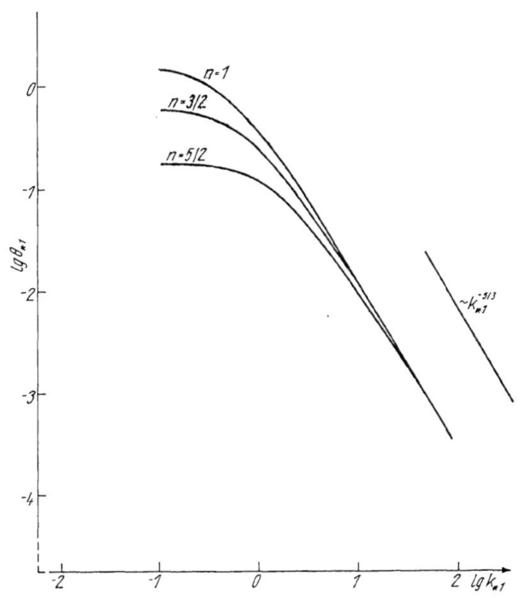

Abb. 4. Eindimensionale Spektren der Turbulenzenergie und der skalaren Schwankungen im Trägheits-Konvektionsbereich $k \gg k_0$

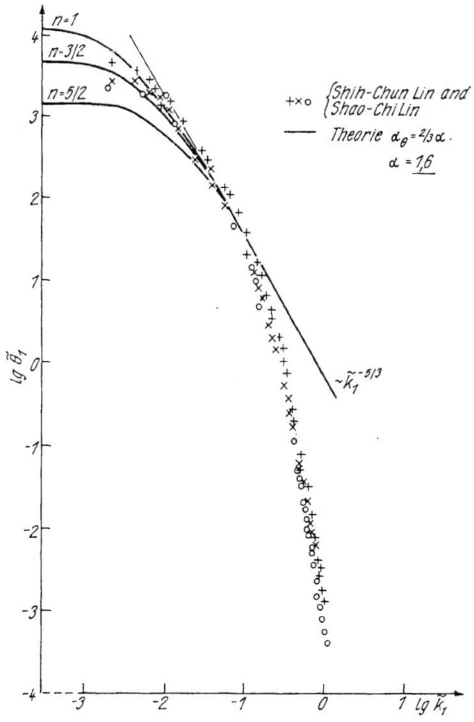

Abb. 5. Eindimensionales Spektrum der skalaren Schwankungen im Trägheitskonvektionsbereich bei $Re\lambda = 150$

Einen Vergleich von Theorie und Experiment ermöglicht die von SHIH-CHUN LIN und SHAO-CHI LIN [10] durchgeführte Messung der Strömung hinter einem beheizten Gitter, die bei der relativ großen REYNOLDSschen Zahl

$$Re_{M_0}\left(=\frac{UM_0}{\nu}\right)=4\cdot 10^5$$

(M_0 Maschenweite, U mittlere Geschwindigkeit)

vorgenommen wurde. Die Modellkonstante Re_λ der Messung betrug

$$Re_\lambda = 150\,.$$

Setzen wir — wie in [1] — die empirische Konstante $\alpha = 1{,}6$, so erhalten wir mit $\widetilde{\Theta}_1 = \Theta_{*1}\alpha_\Theta M^{-5/2}$, $\widetilde{k}_1 = k_{*1}M^{3/2}$ nach (64), (57) und $\alpha_\Theta = \frac{2}{3}\alpha$ nach (54) bzw. $\alpha_{\Theta_1} = 0{,}64$ nach (56) die in Abb. 5 aufgetragenen Verteilungen für $n = 1$, $n = 3/2$ und $n = 5/2$.

Zu bemerken ist, daß $\alpha_{\Theta_1} = 0{,}64$ sich wenig von dem von LIN/LIN aus der Messung abgelesenen Wert $\alpha_{\Theta_1} = 0{,}60$ unterscheidet. Es ist ersichtlich, daß wir im Rahmen der Messung sowohl für $n = 1$ und $n = 3/2$ eine gute Beschreibung des gemessenen eindimensionalen Spektrums im Trägheits-Konvektionsbereich erhalten, wobei hier die erstere Verteilung die Messung anscheinend besser wiedergibt. Dagegen erhält man für $n = 5/2$ (Messung LIN/LIN) keine Beschreibung des eindimensionalen Spektrums im genannten Bereich.

Literatur

1 SZABLEWSKI, W., Ähnlichkeitsgesetz des Spektrums der Turbulenzenergie für den Trägheitsbereich $k \gg k_0$ in homogener isotroper Turbulenz, ZAMM **58**, 143 (1978).
2 SCHEDVIN, J., STEGEN, G. R. and GIBSON, C. H., Universal similarity at high grid REYNOLDS numbers, Fluid Mech. **65**, 561 (1974).
3 HINZE, J. O., Turbulence, McGraw-Hill, New York 1975.
4 KOLMOGOROV, A. N., Die lokale Struktur der Turbulenz in einer inkompressiblen zähen Flüssigkeit bei sehr großen Reynoldsschen Zahlen, Dokl. Akad. Wiss. USSR **30**, 745 (1941).
5 OBUCHOW, A. M., Die Struktur des Temperaturfeldes einer turbulenten Strömung, Izvestija AN SSSR, Ser. Geograph. Geophys. **13**, 58 (1949).
6 CORRSIN, S., On the spectrum of isotropic temperature fluctuations in an isotropic turbulence, J. Appl. Phys. **23**, 469 (1951).
7 KOVASZNAY, L. S. G., Spectrum of locally isotropic turbulence, J. Aeron. Sci. **15**, 745 (1948).
8 MILLS, R. R., KISTLER, A. L., O'BRIEN, V. and CORRSIN, S., NACA TN-4288 (1959).
9 YEH, T. T. and VAN ATTA, C. W., Spectral transfer of scalar and velocity fields in heated-grid turbulence, Fluid Mech. **58**, 233 (1973).
10 SHIH-CHUN LIN and SHAO-CHI LIN, Study of strong temperature mixing in subsonic grid turbulence, Phys. Fluids **16**, 1587 (1973).
11 GIBSON, C. H. and SCHWARZ, W. H., Detection of conductivity fluctuations in a turbulent field, Fluid Mech. **16**, 357 (1963).
12 GIBSON, C. H., STEGEN, G. R. and WILLIAMS, R. B., Statistics of the fine structure of turbulent velocity and temperature fields measured at high REYNOLDS number, Fluid Mech. **41**, 153 (1970).
13 BOSTON, N. E. J. and BURLING, R. W., An investigation of high-wavenumber temperature and velocity spectra in air, Fluid Mech. **55**, 473 (1972).
14 WILLIAMS, R. M. and PAULSON, C. A., Microscale temperature and velocity spectra in the atmospheric boundary layer, Fluid Mech. **83**, 547 (1977).
15 GRANT, H. L., HUGHES, B. A., VOGEL, W. M. and MOILLIET, A., Statistics of the fine structure of turbulent velocity and temperature fields measured at high REYNOLDS number, Fluid. Mech. **41**, 423 (1968).
16 SZABLEWSKI, W., Zur spektralen Verteilung von Temperatur- und Konzentrationsschwankungen in turbulenten Strömungen, Ann. Physik **35**, 19 (1978).
17 CORRSIN, S., The reactant concentration spectrum in turbulent mixing with a first-order reacting, Fluid. Mech. **11**, 407 (1961).

Eingereicht am 25. 7. 78, in revidierter Fassung am 12. 7. 79

Anschrift: Prof. Dr. W. SZABLEWSKI, Zentralinstitut für Mathematik und Mechanik der Akademie der Wissenschaften der DDR, 1080 Berlin, Mohrenstr. 39

ZAMM 60, 99 –107 (1980)

H. IMMICH

Zweiphasen-Kontinuumstheorie einer Suspension starrer, kugelförmiger Teilchen unter Berücksichtigung der Teilchenrotation *)

Teil I: Volumenmittelung der Bilanzgleichungen

Für eine Suspension starrer, kugelförmiger Teilchen in einem viskosen Fluid wird eine Zweiphasen-Kontinuumstheorie (Zwei-Fluid-Modell) durch Anwenden einer Volumenmittelung hergeleitet. Es wird gezeigt, daß die volumengemittelten Spannungstensoren der kontinuierlichen und der dispersen Phase nicht symmetrisch sind, wenn man die Rotationsträgheit der starren Teilchen und das Wechselwirkungsmoment zwischen den Phasen berücksichtigt. Außerdem treten als Folge der Mittelung Momentenspannungen in der dispersen Phase auf. Teil I enthält die Ausgangsgleichungen und das Verfahren der Volumenmittelung.

For a suspension of rigid, spherical particles in a viscous fluid a two-phase theory (two-fluidmodel) is derived by use of a volume-averaging method. It is shown that by considering the rotational inertia of the solid particles and by taking into account the interphase moment the volume-averaged stress tensors of the continuous and the dispersed phase both are not symmetric. Moment stresses arise in the dispersed phase as a result of volume-averaging. Part I contains the governing equations and the procedure of volume-averaging.

Для суспензии, состоящей из абсолютно твёрдых шарообразных частичек в вязкой жидкости, с использованием усреднения по объёму строится теория двухфазного континуума. Показывается, что усреднённые по объёму тензора напряжений непрерывной и дисперсной фаз не симметричны, если учитывать инерцию вращения твёрдых частичек и момент взаимодействия между фазами. Кроме того, в качестве следствия усреднения в дисперсной фазе появляются моментные напряжения. Часть I включает в себя исходные уравнения и метод объёмного усреднения.

1. Einleitung

Eine theoretische Beschreibung einer Zweiphasenströmung, die aus einer kontinuierlichen Phase (Flüssigkeit oder Gas) und einer dispersen Phase (feste Teilchen, Tropfen oder Blasen) besteht, muß in der Lage sein, die beobachtete makroskopische Bewegung einer jeden Phase wiederzugeben. Dabei sind zwei unterschiedliche Aufgaben zu lösen. Die erste Aufgabe besteht in der Verallgemeinerung der bekannten Bilanzgleichungen der Strömungsmechanik in der Weise, daß die Bewegung einer jeden Phase beschrieben wird. Die zweite Aufgabe beinhaltet die Formulierung von geeigneten Materialgleichungen (rheologische Zustandsgleichungen, constitutive equations), welche die Materialeigenschaften berücksichtigen und unbekannte Größen (z. B. Spannungstensoren) als Funktionen der Variablen und der physikalischen Parameter darstellen.

Die vorliegende Untersuchung befaßt sich mit der ersten Aufgabe, d. h. mit der Herleitung von geeigneten Bilanzgleichungen für Masse, Impuls, Impulsmoment und Energie einer jeden Phase unter Verwendung einer Zweiphasen-Kontinuumstheorie.

Bei einer Zweiphasen-Kontinuumstheorie geht man von der Vorstellung aus, auch die disperse Phase als Kontinuum aufzufassen, sofern der mittlere Abstand zwischen den Teilchen der dispersen Phase klein gegenüber einer makroskopischen Abmessung des Strömungsfeldes ist. Die beiden Phasen (kontinuierliche und disperse Phase) werden somit als zwei einander durchdringende Kontinua betrachtet, die miteinander in Wechselwirkung stehen.

Viele Autoren gehen bei der Formulierung der Bilanzgleichungen für eine Zweiphasen-Kontinuumstheorie von einer Analogie zu der Theorie der einander durchdringenden Kontinua [23] aus. Die auftretenden Variablen, wie Geschwindigkeit, Druck und Temperatur sind mittlere Größen, deren Definition für jede Phase vorausgesetzt wird. Beispiele hierzu findet man in [9], [13], [17], [18], [20], [22] und [25]. Dabei bleiben viele offene Fragen, die zu einer Vielzahl widersprüchlicher Bilanzgleichungen geführt haben, die sich vor allem in den Definitionen der Drücke und der Spannungstensoren voneinander unterscheiden.

Eine weitere Möglichkeit der Herleitung einer Zweiphasen-Kontinuumstheorie besteht in einer Mittelung der Bilanzgleichungen, die für die lokalen Variablen einer jeden Phase aufgestellt werden.

Die Mittelung kann dabei über ein bestimmtes Zeitintervall oder über ein Volumen, das eine repräsentative Zahl von Teilchen enthält, erfolgen. Die zeitliche Mittelung wurde vor allem von ISHII [15] untersucht, Beispiele für eine Volumenmittelung findet man in [4], [14], [21] und [24]. Beide Mittelungen eignen sich dazu, Schwankungen der lokalen Variablen (z. B. Abweichung der lokalen Geschwindigkeit im Fluid von der mittleren Geschwindigkeit der kontinuierlichen Phase infolge der Umströmung eines Teilchens oder gegebenenfalls durch Turbulenz) zu glätten, wobei die statistischen Eigenschaften dieser Schwankungen in den gemittelten Bilanzgleichungen berücksichtigt werden (z. B. in den „REYNOLDS"-Spannungen). Das von DREW [10], [11] vorgeschlagene Mittelungsverfahren stellt eine Kombination aus zeitlicher und räumlicher Mittelung dar.

In der vorliegenden Untersuchung wird eine Volumenmittelung auf ein Zweiphasengemisch angewendet, um aus den integralen Bilanzgleichungen für die lokalen Variablen einer jeden Phase entsprechende Bilanzgleichungen für die mittleren Variablen zu erhalten. Die kontinuierliche Phase ist dabei ein newtonsches Fluid, die disperse Phase besteht aus starren, kugelförmigen Teilchen. Es wird gezeigt, daß bei Berücksichtigung der Teilchenrotation und der Rotationsträgheit der Teilchen die mittleren Spannungstensoren beider Phasen nicht symmetrisch sind. Ein Vergleich der volumengemittelten Bilanzgleichungen der dispersen Phase mit den entsprechenden Gleichungen für polare Flüssigkeiten [5], [12] zeigt die enge Verwandtschaft der volumengemittelten Gleichungen der dispersen Phase mit einem polaren Fluid.

*) Auszug aus einer vom Fachbereich für Maschinenwesen der TU München am 13. 2. 1979 angenommenen Dissertation [14] am Lehrstuhl für Strömungsmechanik der TU München

2. Bilanzgleichungen für die lokalen Variablen in integraler, phasengetrennter Form

2.1 Materielles Volumen

Für ein beliebiges Zweiphasengemisch ist ein materielles (mitbewegtes) Volumen $V(t)$ in Bild 1 dargestellt. Die Oberfläche des Volumens $V(t)$ sei $A(t)$, die Summe der Phasengrenzflächen in V, welche die kontinuierliche von der dispersen Phase trennen, werde mit $S(t)$ bezeichnet. Falls Massenstrom durch die Phasengrenzfläche infolge Phasenübergang auftritt, ist $S(t)$ nicht materiell. Das bedeutet, daß die Normalgeschwindigkeitskomponenten von $S(t)$ und der angrenzenden beiden Phasen verschieden sein können.

Die Phasengrenzfläche $S(t)$ kann zunächst endliche Dicke und eine zugehörige Masse besitzen, außerdem seien Grenzflächenspannungen in $S(t)$ zugelassen. Da Teile von $A(t)$ mit $S(t)$ zusammenfallen können, vgl. Bild 1, sei vereinbart, daß in diesem Fall $S(t)$ außerhalb von $V(t)$ liege. Damit ist $A(t)$ überall materiell und bewegt sich mit der lokalen Geschwindigkeit der kontinuierlichen bzw. dispersen Phase.

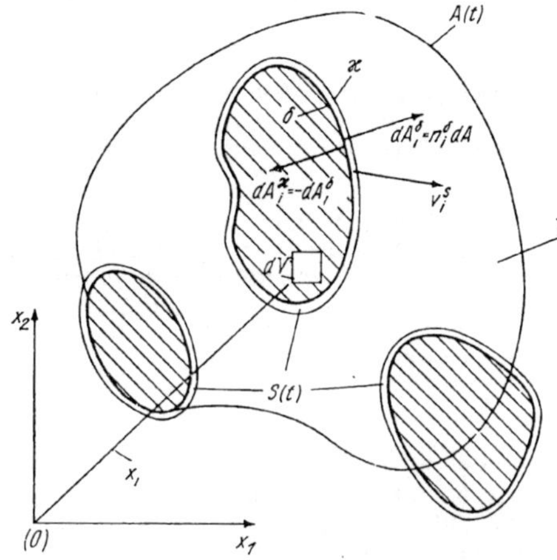

Bild 1. Materielles Volumen in einem Zweiphasengemisch

Mit diesen Vereinbarungen kann man die Bilanzgleichungen für Masse, Impuls, Impulsmoment und Energie für ein Kontinuum [1], [11] in der gleichen Form auch auf das materielle Volumen $V(t)$ anwenden, wenn man beachtet, daß die Variablen in den einzelnen Phasen verschiedene Werte annehmen, vgl. [10], [11].

2.2 Phasentrennung

Bei einer Zweiphasentheorie soll die Bewegung jeder Phase für sich unter dem Einfluß der anderen beschrieben werden. Hierzu benötigt man zum einen die Bilanzgleichungen für den Teil von $V(t)$, der nur die kontinuierliche Phase enthält und zum anderen für den Teil, der nur die disperse Phase erfaßt. Durch Einführen einer Phasenfunktion $\tilde{\alpha}(x_i, t)$ kann diese Phasentrennung vorgenommen werden:

$$\tilde{\alpha}(x_i, t) = \begin{cases} 1 & \text{für} \quad x_i \in d^{*)} \\ 0 & \text{für} \quad x_i \in c\,. \end{cases} \tag{1}$$

Damit erhält man folgende Beziehungen:

$$V_{\mathrm{disp}}(t) = \int_{V(t)} \tilde{\alpha}\, \mathrm{d}V\,, \qquad V_{\mathrm{cont}}(t) = \int_{V(t)} (1 - \tilde{\alpha})\, \mathrm{d}V\,. \tag{2a, b}$$

Dabei bezeichnet $V_{\mathrm{disp}}(t)$ den Teil von $V(t)$, der innerhalb der dispersen Phase liegt und $V_{\mathrm{cont}}(t)$ den Teil, der innerhalb der kontinuierlichen Phase liegt. Die Phasengrenzfläche $S(t)$, die endliche Dicke besitzen kann, befindet sich damit außerhalb von $V_{\mathrm{cont}}(t)$ und $V_{\mathrm{disp}}(t)$. Bei Phasenübergang sind die Teilvolumina $V_{\mathrm{disp}}(t)$ und $V_{\mathrm{cont}}(t)$ nicht materiell, da Teile der die Volumina $V_{\mathrm{disp}}(t)$ und $V_{\mathrm{cont}}(t)$ begrenzenden Oberflächen von der nicht materiellen Phasengrenzfläche $S(t)$ gebildet werden.

2.3 Bilanzgleichungen

Aus den Bilanzgleichungen für ein Kontinuum kann man nach Einführen der Phasenfunktion $\tilde{\alpha}$ und durch Anwenden des Transporttheorems für die nicht materiellen Volumina $V_{\mathrm{disp}}(t)$ und $V_{\mathrm{cont}}(t)$, vgl. [2], die Bilanzgleichungen für jede Phase ableiten [11]. Dabei sind alle auftretenden Variablen lokale, d. h. nicht gemittelte Größen.

a) Massenbilanz

disperse Phase

$$\frac{\mathrm{d}}{\mathrm{d}t} \int_{V(t)} \tilde{\alpha} \varrho\, \mathrm{d}V = - \int_{S(t)} \varrho_\delta (v_i^\delta - v_i^S)\, \mathrm{d}A_i^\delta\,, \tag{3a}$$

kontinuierliche Phase

$$\frac{\mathrm{d}}{\mathrm{d}t} \int_{V(t)} (1 - \tilde{\alpha}) \varrho\, \mathrm{d}V = - \int_{S(t)} \varrho_\varkappa (v_i^\varkappa - v_i^S)\, \mathrm{d}A_i^\varkappa\,. \tag{3b}$$

---

[1]) $x_i \in d$ bzw. $x_i \in c$ bedeutet, daß x_i nur in der dispersen bzw. kontinuierlichen Phase liegen möge.

Dabei bezeichnet ϱ_δ den Randwert der lokalen Dichte ϱ, wenn man sich von der dispersen Phase der Grenzfläche $S(t)$ nähert. Die Integrale auf den rechten Seiten liefern nur dann einen Beitrag, wenn infolge Phasenübergang die lokale Geschwindigkeit v_i^S der Phasengrenzfläche von dem an $S(t)$ angrenzenden Wert der lokalen Geschwindigkeit der dispersen Phase v_i^δ bzw. der kontinuierlichen Phase $v_i^\varkappa$ verschieden ist. Die Richtungen der Oberflächenelemente $\mathrm{d}A_i^\delta$ und $\mathrm{d}A_i^\varkappa$ sind in Bild 1 angegeben.

b) Impulsbilanz

disperse Phase

$$\frac{\mathrm{d}}{\mathrm{d}t}\int\limits_{V(t)} \tilde{\alpha}\varrho v_i\,\mathrm{d}V = \int\limits_A \tilde{\alpha}\pi_{ji}\,\mathrm{d}A_j + \int\limits_V \tilde{\alpha}\varrho f_i\,\mathrm{d}V + \int\limits_S \pi_{ji}^\delta\,\mathrm{d}A_j^\delta - \int\limits_S \varrho_\delta v_i^\delta(v_j^\delta - v_j^S)\,\mathrm{d}A_j^\delta\,, \tag{4a}$$

kontinuierliche Phase

$$\frac{\mathrm{d}}{\mathrm{d}t}\int\limits_{V(t)} (1-\tilde{\alpha})\,\varrho v_i\,\mathrm{d}V = \int\limits_A (1-\tilde{\alpha})\,\pi_{ji}\,\mathrm{d}A_j + \int\limits_V (1-\tilde{\alpha})\,\varrho f_i\,\mathrm{d}V + \int\limits_S \pi_{ji}^\varkappa\,\mathrm{d}A_j^\varkappa - \int\limits_S \varrho^\varkappa v_i^\varkappa(v_j^\varkappa - v_j^S)\,\mathrm{d}A_j^\varkappa\,. \tag{4b}$$

Da die Oberfläche $A(t)$ im allgemeinsten Fall Teilchen der dispersen Phase durchschneidet, vgl. Bild 1, enthält der erste Term auf der rechten Seite von Gl. (4a) den Tensor π_{ji} der lokalen Druck- und Reibungsspannungen der dispersen Phase. Dieser Term wird in den Arbeiten von DREW [11], PANTON [21], CROWE [6] und WHITAKER [24] berücksichtigt, während er in den meisten übrigen Arbeiten nicht enthalten ist, vgl. z. B. BUYEVICH [4], MARBLE [17], NIGMATULIN [19], KALININ [16].

Der zweite Term auf den rechten Seiten der Gleichungen (4 a, b) enthält die Massenkraft f_i. Die Integration über die an den Phasengrenzflächen liegenden lokalen Spannungen π_{ji}^δ bzw. $\pi_{ji}^\varkappa$ in dem dritten Term der rechten Seiten beschreibt die Wechselwirkungskraft zwischen beiden Phasen.

Der letzte Term in Gl. (4a) liefert den Impulsstrom von der kontinuierlichen Phase zur dispersen Phase, falls Phasenübergang auftritt.

c) Impulsmomentenbilanz

disperse Phase

$$\frac{\mathrm{d}}{\mathrm{d}t}\int\limits_{V(t)} \tilde{\alpha}\varrho\varepsilon_{ijk}x_j v_k\,\mathrm{d}V = \int\limits_A \tilde{\alpha}\varepsilon_{ijk}x_j\pi_{lk}\,\mathrm{d}A_l + \int\limits_V \tilde{\alpha}\varrho\varepsilon_{ijk}x_j f_k\,\mathrm{d}V + \int\limits_S \varepsilon_{ijk}x_j\pi_{lk}^\delta\,\mathrm{d}A_l^\delta - \int\limits_S \varrho_\delta\varepsilon_{ijk}x_j v_k^\delta(v_l^\delta - v_l^S)\,\mathrm{d}A_l^\delta\,, \tag{5a}$$

kontinuierliche Phase

$$\frac{\mathrm{d}}{\mathrm{d}t}\int\limits_{V(t)} (1-\tilde{\alpha})\,\varrho\varepsilon_{ijk}x_j v_k\,\mathrm{d}V = \int\limits_A (1-\tilde{\alpha})\,\varepsilon_{ijk}x_j\pi_{lk}\,\mathrm{d}A_l + \int\limits_V (1-\tilde{\alpha})\,\varrho\varepsilon_{ijk}x_j f_k\,\mathrm{d}V + $$
$$+\int\limits_S \varepsilon_{ijk}x_j\pi_{lk}^\varkappa\,\mathrm{d}A_l^\varkappa - \int\limits_S \varrho_\varkappa\varepsilon_{ijk}x_j v_k^\varkappa(v_l^\varkappa - v_l^S)\,\mathrm{d}A_l^\varkappa\,. \tag{5b}$$

Dabei bezeichnet der Ausdruck $\tilde{\alpha}\varrho\varepsilon_{ijk}x_j v_k\,\mathrm{d}V$ das Impulsmoment eines differentiell kleinen Volumenelements eines Teilchens der dispersen Phasen bezüglich des Koordinatenursprungs (O), vgl. Bild 1.

d) Energiebilanz

disperse Phase

$$\frac{\mathrm{d}}{\mathrm{d}t}\int\limits_{V(t)} \tilde{\alpha}\varrho\left(u + \frac{1}{2}v_i v_i\right)\mathrm{d}V = \int\limits_A \tilde{\alpha}(\pi_{ji}v_i + q_j)\,\mathrm{d}A_j + \int\limits_V \tilde{\alpha}\varrho(r + f_i v_i)\,\mathrm{d}V + $$
$$+\int\limits_S (q_i^\delta + \pi_{ji}^\delta v_j^\delta)\,\mathrm{d}A_i^\delta - \int\limits_S \varrho_\delta\left(u_\delta + \frac{1}{2}v_i^\delta v_i^\delta\right)(v_j^\delta - v_j^S)\,\mathrm{d}A_j^\delta\,, \tag{6a}$$

kontinuierliche Phase

$$\frac{\mathrm{d}}{\mathrm{d}t}\int\limits_{V(t)} (1-\tilde{\alpha})\,\varrho\left(u + \frac{1}{2}v_i v_i\right)\mathrm{d}V = \int\limits_A (1-\tilde{\alpha})\,(\pi_{ji}v_i + q_i)\,\mathrm{d}A_i + \int\limits_V (1-\tilde{\alpha})\,\varrho(r + f_i v_i)\,\mathrm{d}V + $$
$$+\int\limits_S (q_i^\varkappa + \pi_{ji}^\varkappa v_j^\varkappa)\,\mathrm{d}A_i^\varkappa - \int\limits_S \varrho_\varkappa\left(u_\varkappa + \frac{1}{2}v_i^\varkappa v_i^\varkappa\right)(v_j^\varkappa - v_j^S)\,\mathrm{d}A_j^\varkappa\,. \tag{6b}$$

Die letzten Ausdrücke in beiden Gleichungen beschreiben den Fluß an spezifischer innerer Energie u und spezifischer kinetischer Energie $(\frac{1}{2})\ v_i v_i$ von der einen zur anderen Phase infolge Phasenübergang. Die Wärmestromdichte wird mit q_i bezeichnet, r ist die Wärmequelldichte.

2.4 Sprungrelationen

Um die Randbedingungen an der Phasengrenzfläche S formulieren zu können, werden noch zusätzlich Beziehungen zwischen den Variablen beiderseits der Diskontinuitätsfläche S benötigt. Diese Beziehungen liefern die Sprungrelationen, vgl. [1].

Durch die Wahl der Phasenfunktion $\tilde{\alpha}$ konnten die Bilanzgleichungen (3) bis (6) bisher ohne Einschränkungen für die Phasengrenzfläche S angegeben werden, da S außerhalb der betrachteten Volumina $V_{\text{disp}}(t)$ bzw. $V_{\text{cont}}(t)$ liegt. Für die Sprungrelationen seien von nun an folgende Vereinfachungen vorausgesetzt:

- die Phasengrenzfläche S sei unendlich dünn, d. h. sie kann mathematisch als Diskontinuitätsfläche angesehen werden,
- die Phasengrenzfläche S besitze keine Masse,
- Grenzflächenspannungen sollen nicht auftreten,
- die Phasengrenzfläche enthalte keine Quellen für Masse, Impuls, Impulsmoment und Energie.

Der allgemeine Fall ohne diese Vereinfachungen ist von DELHAYE [8] untersucht worden.

Die Sprungrelationen lauten somit:

a) Sprungrelation für die Masse

$$[\{\varrho v_i\} - \{\varrho\}\, v_i^S]\, n_i^\delta = 0\,. \tag{7a}$$

Die Klammern { } bezeichnen dabei den Sprung der entsprechenden Variablen beim Durchgang durch die Phasengrenzfläche S in Richtung n_i^δ (vgl. Bild 1):

$$\{\varrho\} = \varrho_\delta - \varrho_\varkappa\,, \qquad \{\varrho v_i\} = \varrho_\delta v_i^\delta - \varrho_\varkappa v_i^\varkappa\,. \tag{7 b, c}$$

b) Sprungrelation für den Impuls:

$$[\{\varrho v_i v_j\} - \{\varrho v_i\}\, v_j^S - \{\pi_{ji}\}]\, n_j^\delta = 0\,. \tag{8}$$

c) Sprungrelation für das Impulsmoment:

$$\varepsilon_{ijk} x_j\, [\{\varrho v_k v_l\} - \{\varrho v_k\}\, v_l^S - \{\pi_{kl}\}]\, n_l^\delta = 0\,. \tag{9}$$

d) Sprungrelation für die Energie:

$$[\{\varrho\,(u + \tfrac{1}{2}\, v_i v_i)\, v_j\} - \{\varrho(u + \tfrac{1}{2}\, v_i v_i)\}\, v_j^S - \{\pi_{ji} v_i + q_j\}]\, n_j^\delta = 0\,. \tag{10}$$

3. Volumenmittelung für eine Suspension kugelförmiger Teilchen

3.1 Volumenmittelung

Eine theoretische Ermittlung der lokalen Variablen bereitet bei praktisch vorkommenden Zweiphasenströmungen außerordentlich große mathematische Schwierigkeiten, da die Lage der zahlreichen Phasengrenzflächen im allgemeinen unbekannt ist. Bei einer ingenieurmäßigen Beschreibung einer Zweiphasenströmung ist man meistens nicht an den Einzelheiten des Strömungsfeldes z. B. in der Umgebung einzelner Teilchen interessiert, vielmehr ist eine Beschreibung der makroskopischen Effekte in gemittelten Variablen in den meisten Fällen ausreichend.

Für eine derartige Betrachtungsweise eignen sich die oben erwähnten Kontinuumsmodelle. Im allgemeinen kann man die disperse Phase nur dann als Kontinuum auffassen, wenn der mittlere Teilchenabstand l klein gegenüber einer charakteristischen Abmessung L des Strömungsfeldes ist:

$$l \ll L\,. \tag{11}$$

Falls die Bedingung $l \ll L$ erfüllt ist, kann man mit einer Volumenmittelung aus der dispersen und der kontinuierlichen Phase zwei einander durchdringende Kontinua herleiten, vgl. [4], [24].

Anstelle einer einfachen Volumenmittelung führt DREW [10], [11] eine zweifache Volumenmittelung und zusätzlich eine zweifache zeitliche Mittelung ein, um räumliche und zeitliche Diskontinuitäten zu glätten und um einmal stetig differenzierbare gemittelte Variable zu erhalten. Wie ISHII [15] erwähnt, erscheint die Einführung dieser vierfachen Mittelung nicht notwendig, um gemittelte Gleichungen für die makroskopische Bewegung zu erhalten. Vielmehr genügt es, entweder eine Volumenmittelung oder eine zeitliche Mittelung allein durchzuführen [15].

Daher wird im folgenden eine einfache Volumenmittelung verwendet, wobei das in Kapitel 3.4 verwendete Verfahren der Volumenmittelung der Bilanzgleichungen eng mit dem von DREW [11] vorgeschlagenen Verfahren verwandt ist.

Ein für eine Volumenmittelung geeignetes raumfestes, kugelförmiges Volumen V_0 mit einem Radius l_0 ist in Bild 2 dargestellt. Der Radius l_0 muß dabei klein gegenüber einer Abmessung L des Strömungsfeldes sein, aber groß

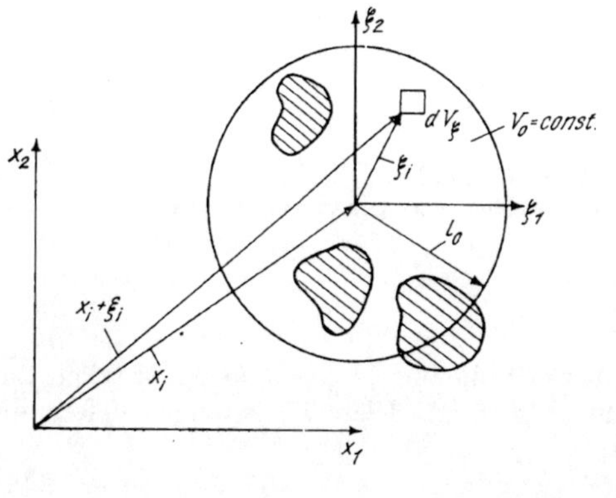

Bild 2. Mittelung über ein kugelförmiges, raumfestes Volumen V_0

genug, um eine repräsentative Zahl von Teilchen zu erfassen. Wie WHITAKER [24] und BRENNER [3] zeigen, muß l_0 im allgemeinen von der gleichen Größenordnung wie der mittlere Teilchenabstand l sein. Damit gilt entsprechend Gl. (11):

$$l_0 \ll L , \qquad l_0 = 0(l) . \tag{12}$$

Für eine lokale Variable $\psi(x_i, t)$ erhält man durch Mittelung über das gesamte Volumen $V = V_0$ den Mittelwert $\langle\ \rangle$:

$$\langle \psi(x_i, t) \rangle = \frac{1}{V_0} \int\limits_{V_0} \psi(x_i + \xi_i, t)\, \mathrm{d}V_\xi . \tag{13}$$

Für ein Kontinuumsmodell, in dem man das Zweiphasengemisch als ein Kontinuum betrachtet, ist dies die geeignete Mittelung, da $\langle\psi\rangle$ durch Mittelung über beide Phasen gebildet wird.

Für die dieser Arbeit zugrundeliegende Zweiphasen-Kontinuumstheorie kann mit Hilfe der Phasenfunktion $\tilde{\alpha}$ der Mittelwert für die disperse Phase definiert werden:

$$\langle \tilde{\alpha}(x_i, t)\, \psi(x_i, t) \rangle = \frac{1}{V_0} \int\limits_{V_0} \tilde{\alpha}(x_i + \xi_i, t)\, \psi(x_i + \xi_i, t)\, \mathrm{d}V \tag{14}$$

und entsprechend für die kontinuierliche Phase:

$$\langle (1 - \tilde{\alpha})\, \psi \rangle = \frac{1}{V_0} \int\limits_{V_0} (1 - \tilde{\alpha})\, \psi\, \mathrm{d}V . \tag{15}$$

Eine wichtige Beziehung erhält man, wenn man in den Gln. (14) und (15) die Variable $\psi \equiv 1$ setzt und die Definition für V_{disp} nach Gl. (2a) beachtet:

$$\langle \tilde{\alpha}(x_i, t) \rangle = \frac{1}{V_0} \int\limits_{V_0} \tilde{\alpha}(x_i + \xi_i, t)\, \mathrm{d}V = \frac{1}{V_0} \sum_{V_P^\nu \text{ in } V_0} V_P^\nu = \frac{V_{\text{disp}}}{V_0} = \alpha(x_i, t) \underset{(-)}{<} 1 . \tag{16}$$

Der Mittelwert über die Phasenfunktion $\tilde{\alpha}$ liefert somit den Volumenanteil $\alpha(x_i, t)$ der dispersen Phase.

Entsprechend erhält man den Volumenanteil der kontinuierlichen Phase mit V_{cont} nach Gl. (2b):

$$\langle 1 - \tilde{\alpha} \rangle = \frac{1}{V_0} \int\limits_{V_0} (1 - \tilde{\alpha})\, \mathrm{d}V = \frac{V_{\text{cont}}}{V_0} = 1 - \alpha \leqq 1 . \tag{17}$$

In Analogie zu DREW [11] führen wir für die gemittelten Variablen folgende Bezeichnung ein:

$$\langle \tilde{\alpha}\psi \rangle = \frac{1}{V_0} \int\limits_{V_0} \tilde{\alpha}\psi\, \mathrm{d}V = \alpha\psi_d , \tag{18}$$

$$\langle (1 - \tilde{\alpha})\, \psi \rangle = \frac{1}{V_0} \int\limits_{V_0} (1 - \tilde{\alpha})\, \psi\, \mathrm{d}V = (1 - \alpha)\, \psi_c . \tag{19}$$

Die Indices d bzw. c bezeichnen den Mittelwert der entsprechenden Variablen für die disperse bzw. kontinuierliche Phase.

Eine wichtige Beziehung erhält man, wenn man den Mittelwert einer Variablen nochmals mittelt:

$$\langle \tilde{\alpha}\psi_d \rangle = \frac{1}{V_0} \int\limits_{V_0} \tilde{\alpha}\psi_d\, \mathrm{d}V \approx \alpha\psi_d , \tag{20}$$

$$\langle (1 - \tilde{\alpha})\, \psi_c \rangle = \frac{1}{V_0} \int\limits_{V_0} (1 - \tilde{\alpha})\, \psi_c\, \mathrm{d}V \approx (1 - \alpha)\, \psi_c . \tag{21}$$

Eine genauere Untersuchung und Begründung findet man in [10]. Falls ψ und χ zwei Variable sind, erhält man mit (18) und (19) die Beziehungen:

$$\langle \tilde{\alpha}(\psi + \chi) \rangle = \alpha\psi_d + \alpha\chi_d , \tag{22}$$

$$\langle (1 - \tilde{\alpha})\, (\psi + \chi) \rangle = (1 - \alpha)\, \psi_c + (1 - \alpha)\, \chi_c . \tag{23}$$

Da einige Mittelwerte durch Mittelung über die Diskontinuitätsfläche S bestimmt werden, wie z. B. die Wechselwirkungskraft, vgl. Gl. (35) und (49 a, b), führen wir in Analogie zu DREW [11] noch eine weitere Mittelung über die Diskontinuitätsfläche S ein:

$$\Phi_d(x_i, t) = \frac{1}{V_0} \int\limits_S \Phi_\delta(x_i + \xi_i, t)\, \mathrm{d}A , \tag{24}$$

$$\Phi_c(x_i, t) = \frac{1}{V_0} \int\limits_S \Phi_\varkappa(x_i + \xi_i, t)\, \mathrm{d}A . \tag{25}$$

Die Integration hat dabei über alle in V_0 liegenden Phasengrenzflächen S zu erfolgen. Setzt man in (24) bzw. (25) Φ_δ bzw. $\Phi_\varkappa \equiv 1$, so erhält man die Summe der Phasengrenzflächen pro Volumen V_0.

3.2 Starre kugelförmige Teilchen

Die Volumenmittelung soll für den Fall einer Suspension starrer kugelförmiger Teilchen durchgeführt werden. Das Zweiphasengemisch bestehe aus folgenden Anteilen:

disperse Phase: starre, kugelförmige, homogene Teilchen gleicher Dichte ϱ_d und vom gleichen Radius a. Die Teilchen seien so groß, daß die BROWNsche Molekülbewegung keinen Einfluß auf ihre Bewegung ausübt.

kontinuierliche Phase: homogenes newtonsches Fluid (konstante Dichte ϱ_c und konstante Viskosität μ).

Die Teilchenrotation wurde bei der Herleitung von volumengemittelten Bilanzgleichungen bisher nur von BUYEVICH [4] berücksichtigt. Da BUYEVICH jedoch ein anderes Verfahren der Volumenmittelung der Bilanzgleichungen verwendet und von vornherein einige einschränkende Vereinfachungen trifft, weichen seine Ergebnisse von den hier erzielten Ergebnissen ab. Die Unterschiede werden in Kapitel 5.1 besprochen.

Phasenübergang wird nicht berücksichtigt. Der Fall mit Phasenübergang wurde in [14] untersucht.

Da bei der Volumenmittelung für die disperse Phase zunächst über die einzelnen Teilchen integriert wird und anschließend diese Mittelwerte über die einzelnen Teilchen über das Volumen V_0 gemittelt werden, werden im folgenden einige z. T. aus der Kinetik der starren Körper bekannte Beziehungen für Kugeln bereitgestellt.

a) Translations- und Rotationsbewegung

Da die disperse Phase aus festen, kugelförmigen Teilchen besteht, wird ihre Bewegung von zwei unabhängigen kinematischen Variablen beschrieben, der Translationsgeschwindigkeit v_i^P eines Teilcheus und seiner Rotationsgeschwindigkeit ω_i.

Die Geschwindigkeit v_i eines Teilvolumens dV eines Teilchens lautet somit (Bild 3):

$$v_i = v_i^P + \varepsilon_{ijk}\omega_j x'_k \,. \tag{26}$$

Das Koordinatensystem x'_i sei ein mit der Translationsgeschwindigkeit v_i^P mitbewegtes Koordinatensystem mit dem Ursprung im Teilchenmittelpunkt = Teilchenschwerpunkt.

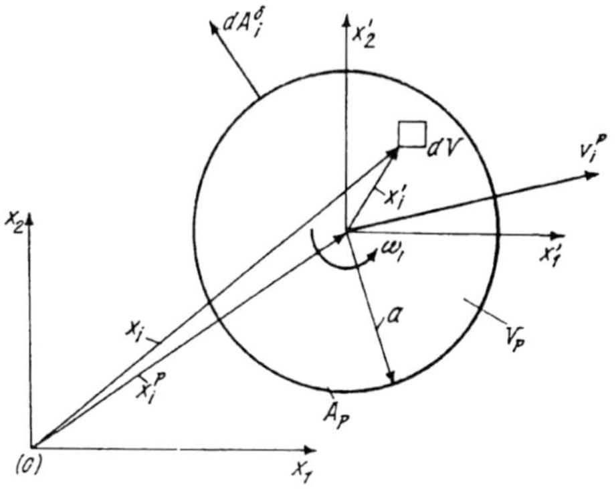

Bild 3. Translation und Rotation eines Teilchens vom Volumen V_P

b) spezifischer Impuls

Die Mittelung des spezifischen Impulses über V_P liefert mit dem Schwerpunktsatz:

$$\frac{1}{V_P}\int_{V_P} v_i \, dV = v_i^P + \varepsilon_{ijk}\omega_j \frac{1}{V_P}\int_{V_P} \overset{0}{x'_k} \, dV = v_i^P \,. \tag{27}$$

c) spezifisches Impulsmoment

Die Mittelung des spezifischen Impulsmoments über V_P ergibt mit Gl. (26):

$$\frac{1}{V_P}\int_{V_P} \varepsilon_{ijk} x_j v_k \, dV = \varepsilon_{ijk} x_j^P v_k^P + \vartheta\omega_i \quad \text{mit} \quad \vartheta = \tfrac{2}{5} a^2 \,. \tag{28}$$

Hierbei bezeichnet ϑ das auf die Teilchenmasse bezogene Trägheitsmoment (Rotationsträgheit) des kugelförmigen Teilchens.

d) spezifische kinetische Energie

Die Integration der spezifischen kinetischen Energie über V_P ergibt:

$$\frac{1}{V_P}\int_{V_P} \frac{1}{2} v_i v_i \, dV = \frac{1}{2} v_i^P v_i^P + \frac{1}{2} \vartheta \omega_i \omega_i \,. \tag{29}$$

e) Spannungsmoment

Die Integration des Moments der innerhalb eines Teilchens auftretenden Spannungen um den Teilchenmittelpunkt liefert:

$$\varepsilon_{ikl} \frac{1}{V_P}\int_{V_P} x'_k \pi_{jl} \, dV = \lambda_{ij}^P \,. \tag{30}$$

Dabei bezeichnet λ_{ij}^P den Tensor der Momentenspannungen. In einer Suspension starrer kugelförmiger Teilchen ist nach der Volumenmittelung nur dann ein wesentlicher Beitrag $\lambda_{ij}^d \neq 0$ zu erwarten, wenn durch direkten Kontakt der Teilchen untereinander eine Spannungsverteilung innerhalb der Teilchen erzeugt wird, die in Gl. (30) einen Beitrag zu λ_{ij}^P liefert. Dies ist im allgemeinen nur für große Werte des Volumenanteils ($\alpha \to \alpha_{max}$) zutreffend.

f) Kraft vom Fluid auf ein Teilchen

In folge der Wechselwirkung mit dem Fluid erfährt das Teilchen vom Fluid die Kraft F_i:

$$F = \int_{A_P} \pi_{ji}^\delta \, dA_j^\delta \,. \tag{31}$$

g) Moment vom Fluid auf ein Teilchen

Das Moment, das vom Fluid auf ein Teilchen ausgeübt wird, lautet:

$$M_i^P = \varepsilon_{ijk} \int_{A_P} x_j' \pi_{kl}^\delta \, \mathrm{d}A_l^\delta \,. \tag{32}$$

In beiden Fällen, Gl. (31) und (32), muß über die in der Teilchenoberfläche liegenden Spannungen π_{ji}^δ integriert werden.

3.3 Verwendete Mittelwerte

Die verwendeten Mittelwerte sind in den folgenden Tabellen zusammengefaßt. Die Mittelungen sind dabei nach den Definitionen (18), (19) bzw. (24) und (25) vorzunehmen, die gemittelten Variablen der dispersen bzw. kontinuierlichen Phase werden mit dem Index d bzw. c bezeichnet.

a) Mittelwerte der lokalen Variablen

Tabelle 1. Mittelwerte der lokalen Variabeln

Variable	disperse Phase	kontinuierliche Phase	Gleichung Nr.
Volumenanteil	$\langle\tilde{\alpha}\rangle = \alpha$	$\langle 1 - \tilde{\alpha}\rangle = 1 - \alpha$	(33a, b)
Dichte	$\langle\tilde{\alpha}\varrho\rangle = \alpha\varrho_d$	$\langle(1 - \tilde{\alpha})\,\varrho\rangle = (1 - \alpha)\,\varrho_c$	(34a, b)
Geschwindigkeit	$\langle\tilde{\alpha}\varrho v_i\rangle = \alpha\varrho_d v_i^d$	$\langle(1 - \tilde{\alpha})\,\varrho v_i\rangle = (1 - \alpha)\,\varrho_c v_i^c$	(35a, b)
Rotation	$\langle\tilde{\alpha}\varrho\vartheta\omega_i\rangle = \alpha\varrho_d\vartheta_d\omega_i^d$	—	(36)
Druck	$\langle\tilde{\alpha}p\rangle = \alpha p_d$	$\langle(1 - \tilde{\alpha})\,p\rangle = (1 - \alpha)\,p_c$	(37a, b)
Spannungen	$\langle\tilde{\alpha}\pi_{ij}\rangle = \alpha\pi_{ij}^d$	$\langle(1 - \tilde{\alpha})\,\pi_{ij}\rangle = (1 - \alpha)\,\pi_{ij}^c$	(38a, b)
Momentenspannungen	$\langle\tilde{\alpha}\lambda_{ij}^P\rangle = \alpha\lambda_{ij}^d$	—	(39)
innere Energie	$\langle\tilde{\alpha}\varrho u\rangle = \alpha\varrho_d u_d$	$\langle(1 - \tilde{\alpha})\,\varrho u\rangle = (1 - \alpha)\,\varrho_c u_c$	(40a, b)
Wärmestromdichte	$\langle\tilde{\alpha}q_i\rangle = \alpha q_i^d$	$\langle(1 - \tilde{\alpha})\,q_i\rangle = (1 - \alpha)\,q_i^c$	(41a, b)
Wärmequelldichte	$\langle\tilde{\alpha}\varrho r\rangle = \alpha\varrho_d r_d$	$\langle(1 - \tilde{\alpha})\,\varrho r\rangle = (1 - \alpha)\,\varrho_c r_c$	(42a, b)

b) diffusive Terme

Zusätzlich zu den Mittelwerten der lokalen Variablen treten bei der Volumenmittelung der Bilanzgleichungen Terme auf, welche die Differenz der lokalen und der mittleren Variablen enthalten. Es ist üblich, die Abweichung der lokalen Variablen von den mittleren Variablen als Diffusion zu bezeichnen, vgl. [11], [23]. Die in der Tabelle 2 aufgeführten Terme werden deshalb als diffusive Terme bezeichnet.

Tabelle 2. Diffusive Terme

Bezeichnung	disperse Phase	kontinuierliche Phase	Gleichung Nr.
diffusiver Impulsstrom	$\langle\tilde{\alpha}\varrho(v_i^P - v_i^d)\,(v_j^P - v_j^d)\rangle = \alpha\varrho_d J_{ij}^d$	$\langle(1 - \tilde{\alpha})\,\varrho(v_i - v_i^c)\,(v_j - v_j^c)\rangle = (1 - \alpha)\,\varrho_c J_{ij}^c$	(43a, b)
diffusiver Impulsmomentenstrom	$\langle\tilde{\alpha}\varrho\vartheta(v_i^P - v_i^d)\,(\omega_j - \omega_j^d)\rangle = \alpha\varrho_d\vartheta_d D_{ij}^d$	—	(44)
diffusive kinetische Energie der Rotationsbewegung	$\langle\tilde{\alpha}\varrho\vartheta(\omega_i - \omega_i^d)\,(\omega_i - \omega_i^d)\rangle = \alpha\varrho_d\vartheta_d\Omega_{ii}^d$	—	(45)
diffusive Arbeit der Spannungen	$\langle\tilde{\alpha}(v_i^P - v_i^d)\,\pi_{ji}n_j\rangle = \alpha P_j^d n_j$	$\langle(1 - \tilde{\alpha})\,(v_i - v_i^c)\,\pi_{ji}n_j\rangle = (1 - \alpha)\,P_j^c n_j$	(46a, b)
diffusive Arbeit der Momentenspannungen	$\langle\tilde{\alpha}(\omega_i - \omega_i^d)\,\lambda_{ji}^P n_j\rangle = \alpha L_j^d n_j$	—	(47)
diffusiver Energiestrom	$\langle\tilde{\alpha}\varrho(u_P - u_d)\,(v_j^P - v_j^d)\rangle + \langle\tilde{\alpha}\varrho\,\frac{1}{2}\,(v_i^P - v_i^d)\,(v_i^P - v_i^d)\,(v_j^P - v_j^d)\rangle + \langle\tilde{\alpha}\varrho\,\frac{1}{2}\,\vartheta(\omega_i - \omega_i^d)\,(\omega_i - \omega_i^d) \times (v_j^P - v_j^d)\rangle = \alpha\varrho_d G_j^d$	$\langle(1 - \tilde{\alpha})\,\varrho(u - v_c)\,(v_j - v_j^c)\rangle + \langle(1 - \tilde{\alpha})\,\varrho\,\frac{1}{2}\,(v_i - v_i^c)\,(v_i - v_i^c) \times (v_j - v_j^c)\rangle = (1 - \alpha)\,\varrho_c G_j^c$	(48)

Der diffusive Impulsstrom $\alpha\varrho_d J_{ij}^d$ bzw. $(1 - \alpha)\,\varrho_c J_{ij}^c$ nach Gl. (43a, b) entspricht den REYNOLDSspannungen, die bei der Beschreibung einer turbulenten Strömung in Mittelwerten erscheinen. Bei Mehrphasenströmungen sind auch bei laminarer Strömung die diffusiven Terme zu berücksichtigen, da die lokalen Geschwindigkeiten v_i^P bzw. v_i im allgemeinen von den mittleren Werten v_i^d bzw. v_i^c verschieden sind (z. B. infolge der Geschwindigkeitsverteilung v_i im Fluid in der Umgebung eines Teilchens). Die auftretenden Geschwindigkeitsdifferenzen $v_i^P - v_i^d$ bzw. $v_i - v_i^c$ werden als Diffusionsgeschwindigkeiten bezeichnet [11], [23].

Der diffusive Impulsmomentenstrom $\alpha\varrho_d\vartheta_d D_{ij}^d$ der dispersen Phase stellt das Analogon zu dem diffusiven Impulsstrom dar. Ein der Definition (44) entsprechender Term erscheint bei der kinetischen Theorie einer Flüssigkeit mit Mikrostruktur von DAHLER und SCRIVEN [7], vgl. auch die Diskussion in Kapitel 5.3.

c) Wechselwirkungsterme

Da die Wechselwirkung zwischen den Phasen an den Phasengrenzflächen stattfindet, hat die Mittelung der entsprechenden Wechselwirkungsterme in Tabelle 3 über die in V_0 liegenden Phasengrenzflächen nach Gl. (24) und (25) zu erfolgen.

Tabelle 3. Wechselwirkungsterme

Bezeichnung	disperse Phase	kontinuierliche Phase	Gleichung Nr.
Wechselwirkungskraft	$\langle \pi_{ji}^{\delta} n_j^{\delta} \rangle = \langle F_i \rangle = F_i^d = -F_i^c$	$\langle \pi_{ji}^{\varkappa} n_j^{\varkappa} \rangle = F_i^c = -F_i^d$	(49a, b)
Wechselwirkungsmoment	$\langle \varepsilon_{ijk} x_j'^{\nu} \pi_{kl}^{\delta} n_l^{\delta} \rangle = \langle M_i^P \rangle = M_i^d = -M_i^c$	$\langle \varepsilon_{ijk} x_j'^{\nu} \pi_{kl}^{\varkappa} n_l^{\varkappa} \rangle = M_i^c = -M_i^d$	(50a, b)
Energieerzeugung infolge Wechselwirkung	$\langle q_j^{\delta} n_j^{\delta} \rangle + \langle (v_i^P - v_i^d) F_i \rangle + \langle (\omega_i - \omega_i^d) M_i^P \rangle = E_d$	$E_c = -E_d$	(51a, b)

Die Wechselwirkungskraft F_i^d ist die Kraft/Volumen, die von der kontinuierlichen Phase auf die disperse Phase ausgeübt wird. Mit Hilfe der Sprungrelationen für den Impuls und das Impulsmoment, Gl. (8) und (9), erhält man das Gesetz actio = reactio:

$$F_i^d = -F_i^c \tag{52}$$

$$M_i^d = -M_i^c . \tag{53}$$

3.4. Verfahren der Volumenmittelung der Bilanzgleichungen

Das in Anlehnung an DREW [10], [11] verwendete Verfahren der Volumenmittelung der Bilanzgleichungen wird im folgenden zunächst allgemein erklärt und anschließend für die Massenbilanz der dispersen Phase durchgeführt.

Dazu betrachten wir das in Bild 1 gezeigte mitbewegte materielle Volumen $V(t)$, dessen Oberfläche sich jeweils mit der lokalen Geschwindigkeit der beiden Phasen bewegt. Wir definieren nun ein mit $V(t)$ deckungsgleiches Volumen $V_d(t)$, dessen Oberfläche sich mit der mittleren Geschwindigkeit v_i^d der dispersen Phase bewegen möge. Entsprechend führen wir ein sich mit der mittleren Geschwindigkeit v_i^c der kontinuierlichen Phase mitbewegendes Volumen $V_c(t)$ ein. Beide Volumina sind also für die entsprechenden gemittelten Variablen materiell.

Wir betrachten zunächst Ausdrücke von der Form

$$\frac{\mathrm{d}}{\mathrm{d}t} \int\limits_{V_d(t)} \alpha \psi_d \,\mathrm{d}V \quad \text{bzw.} \quad \frac{\mathrm{d}}{\mathrm{d}t} \int\limits_{V_c(t)} (1-\alpha)\, \psi_c \,\mathrm{d}V ,$$

wobei $\psi_d(x_i, t)$ und $\psi_c(x_i, t)$ gemittelte Größen (z. B. Dichte, Impuls, Impulsmoment, Energie) bezeichnen. Unter Beachtung von Gl. (14) erhält man für den ersten Ausdruck

$$\frac{\mathrm{d}}{\mathrm{d}t} \int\limits_{V_d(t)} \alpha(x_i, t)\, \psi_d(x_i, t) \,\mathrm{d}V = \frac{1}{V_0} \frac{\mathrm{d}}{\mathrm{d}t} \int\limits_{V_d(t)} \int\limits_{V_0} \tilde{\alpha}(x_i + \xi_i, t)\, \psi(x_i + \xi_i, t) \,\mathrm{d}V_\xi \,\mathrm{d}V_x . \tag{54}$$

Dabei bezeichnet V_0 das Kugelvolumen, über das gemittelt wird, mit dem Mittelpunkt in x_i, $\mathrm{d}V = \mathrm{d}V_x$ und $\mathrm{d}V_\xi$ sind Volumenelemente an den Stellen x_i bzw. ξ_i, vgl. Bild 4.

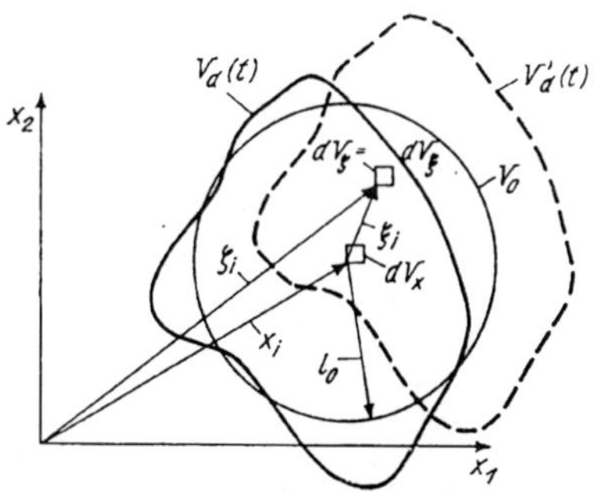

Bild 4. Zur Volumenmittelung über V_0; materielles Volumen für die gemittelten Variablen $V_d(t)$ und dessen Verschiebung $V_d'(t)$ unter der Transformation $x_i + \xi_i = \zeta_i$

Als nächster Schritt wird die Reihenfolge der Integrationen in Gl. (54) vertauscht und die neue Variable $\zeta_i = x_i + \xi_i$ eingeführt:

$$\begin{aligned} \frac{\mathrm{d}}{\mathrm{d}t} \int\limits_{V_d(t)} \alpha \psi_d \,\mathrm{d}V &= \frac{1}{V_0} \int\limits_{V_0} \frac{\mathrm{d}}{\mathrm{d}t} \int\limits_{V_d(t)} \tilde{\alpha}(x_i + \xi_i, t)\, \psi(x_i + \xi_i, t) \,\mathrm{d}V_x \,\mathrm{d}V_\xi \\ &= \frac{1}{V_0} \int\limits_{V_0} \frac{\mathrm{d}}{\mathrm{d}t} \int\limits_{V_d'(t)} \tilde{\alpha}(\zeta_i, t)\, \psi(\zeta_i, t) \,\mathrm{d}V_\zeta \,\mathrm{d}V_\xi . \end{aligned} \tag{55}$$

Dabei bedeutet $V_d'(t)$ die Translation (Verschiebung) von $V_d(t)$ infolge der Transformation ζ_i, vgl. Bild 4. Die Vertauschung der Reihenfolge der Integrationen und die Transformation ζ_i haben zur Folge, daß zunächst die lokalen Variablen $\tilde{\alpha}\psi$ über das durch die Transformation ζ_i verschobene Volumen $V_d'(t)$ integriert werden und die zeitliche Änderung dieses Ausdrucks anschließend über das Kugelvolumen mit dem Ursprung in x_i gemittelt wird.

Die Volumina $V_d(t)$ und $V_d'(t)$ in Gl. (55) sind infolge der Vertauschung der Reihenfolge der Integrationen nicht mehr materiell, da die Oberflächen von $V_d(t)$ und $V_d'(t)$ sich mit der mittleren Geschwindigkeit v_i^d bewegen, die Variable ψ aber eine lokale (nicht gemittelte) Größe darstellt.

Da die in Kapitel 2 bereitgestellten Bilanzgleichungen für ein materielles Volumen $V(t)$ angegeben sind, wird mit Hilfe des Transporttheorems für allgemein bewegte, nicht materielle Volumina [2] der Ausdruck

$$\frac{\mathrm{d}}{\mathrm{d}t} \int\limits_{V_d'(t)} \tilde{\alpha} \psi \,\mathrm{d}V$$

in Gl. (55) in ein Integral über ein materielles Volumen umgeformt. Wir bezeichnen das Volumen, das mit $V_d'(t)$ deckungsgleich ist, sich jedoch mit der lokalen Geschwindigkeit der dispersen Phase bewegt, mit $V_m(t)$, seine Oberfläche mit $A_m(t)$ und erhalten mit Hilfe des Transporttheorems:

$$\frac{\mathrm{d}}{\mathrm{d}t} \int\limits_{V_d'(t)} \tilde{\alpha}(\zeta_i, t)\, \psi(\zeta_i, t)\, \mathrm{d}V = \frac{\mathrm{d}}{\mathrm{d}t} \int\limits_{V_m(t)} \tilde{\alpha}\psi\, \mathrm{d}V - \int\limits_{A_m} \tilde{\alpha}\psi(v_i^P - v_i^d)\, \mathrm{d}A_i^m \,. \tag{56}$$

Der erste Ausdruck auf der rechten Seite von Gl. (56) kann jetzt durch die rechten Seiten der Bilanzgleichungen für die disperse Phase von Kapitel 2 ersetzt werden.

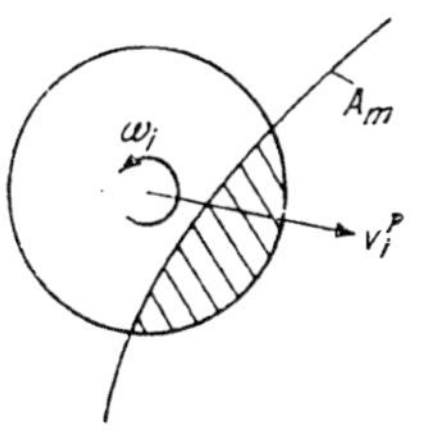

Bild 5. In das Volumen $V_m(t)$ eintretendes, rotierendes Teilchen

Im zweiten Ausdruck auf der rechten Seite spielt nur die Differenz der lokalen Translationsgeschwindigkeit v_i^P zur mittleren Translationsgeschwindigkeit v_i^d eine Rolle, die Teilchenrotation hat keinen Einfluß. Man sieht dies auch ohne Beweis anhand der Skizze in Bild 5 ein:

Da keine Zeitableitung vor dem Integral über A_m in Gl. (56) steht, kann A_m als raumfest aufgefaßt werden. Der Ausdruck $\tilde{\alpha}v_i\, \mathrm{d}A_i^m$ beschreibt das durch A_m eintretende Teilchenvolumen pro Zeit. Wie aus Bild 5 ersichtlich ist, trägt die Drehung des kugelförmigen Teilchens nicht zu einer Änderung des schraffierten Volumens bei. Nur die Translationsgeschwindigkeit v_i^P liefert einen Beitrag. Damit erhält man $\tilde{\alpha}v_i\, \mathrm{d}A_i^m = \tilde{\alpha}v_i^P\, \mathrm{d}A_i^m$.

Literatur

1 BECKER, E. und BÜRGER, W., Kontinuumsmechanik, Stuttgart, Teubner (1975).
2 BIRD, R. B., STEWART, W. E. and LIGHTFOOT, E. N., Transport phenomena, New York, Wiley (1960).
3 BRENNER, H., Rheology of two-phase systems, Ann. Rev. Fluid Mech., Vol. 2, 137–176 (1970).
4 BUYEVICH, YU, A. and MARKOV, V. G., The continuum mechanics of monodisperse suspensions, The integral and differential conservation laws, Int. Chem. Engng., Vol. 15, 687–694 (1975).
5 COWIN, S. C., Polar fluids, Phys. Fluids, Vol. 11, 1919–1927 (1968).
6 CROWE, C. T., Conservation equations for vapor-droplet flows, Proc. 1976 Heat Transfer and Fluid Mech. Inst., Univ. Calif., Davis, June 21–23, 214–228 (1976), Stanf. Univ. Press.
7 DAHLER, J. S. and SCRIVEN, L. E., Theory of structured continua, I. General consideration of angular momentum and polarization, Proc. Roy. Soc., Vol. A275, 504–527 (1963).
8 DELHAYE, J. M., Jump conditions and entropy sources in two-phase systems, Local instant formulation, Int. J. Multiphase Flow, Vol. 1, 395–409 (1974).
9 DEEMTER VAN, J. J. and LAAN VAN DER, E. T., Momentum and energy balances for dispersed two-phase flow, Appl. Sci. Res., Vol. A10, 102–108 (1961).
10 DREW, D. A., Derivation and application of average equations for two-phase media, Ph. D. Thesis, June 1970, Rensselaer Polytechnic Institute, Troy, N.Y.
11 DREW, D. A., Averaged field equations for two-phase media, Stud. Appl. Math., Vol. L, 133–166 (1971).
12 ERINGEN, C. A., Theory of micropolar fluids, J. Math. Mech., Vol. 16, 1–18 (1966).
13 GREEN, A. E. and NAGHDI, P. M., A dynamical theory of interacting continua, Int. J. Engng. Sci. Vol. 3, 231–241 (1965).
14 IMMICH, H., Strömungsverhalten einer Suspension kugelförmiger Teilchen bei plötzlich bewegter unendlich langer Platte nach einer Zweiphasen-Kontinuumstheorie, Diss. TU München (1979).
15 ISHII, M., Thermo-fluid dynamic theory of two-phase flow, Paris, Eyrolles (1975).
16 KALININ, A. V., Derivation of fluid-mechanics equations for a two-phase medium with phase changes, Heat Transfer-Soviet Research, Vol. 2, No. 3, 83–96 (1970).
17 MARBLE, F. E., Dynamics of dusty gases, Ann. Rev. Fluid Mech. Vol. 2, 397–446 (1970).
18 MURRAY, J. D., On the mathematics of fluidization, pt. 1, fundamental equations and wave propagation, J. Fluid Mech., Vol. 21, 465–493 (1965).
19 NIGMATULIN, R. J., Equations of hydromechanics and compression shock in two-velocity and two-temperature continuum with phase-transformations, Izv. AN SSR, Mekhanika Zhidkosti i Gaza, Vol. 2, 33–47 (1967).
20 PAI, S. I., A critical review of the fundamental equations of a mixture of a gas and small solid particles, Z. Flugwiss. Vol. 19, 353–360 (1971).
21 PANTON, R., Flow properties for the continuum viewpoint of a non-equilibrium gas-particle-mixture, J. Fluid Mech., Vol. 31, 273–303 (1968).
22 SOO, S. L., Fluid dynamics of multiphase systems, Blaisdell Publ. Co., Waltham/Mass.-Toronto-London (1967).
23 TRUESDELL, C. and TOUPIN, R. A., The classical field theories in: Handbuch der Physik, Bd. III/1, FLÜGGE, S. (Hsg.), Berlin, Springer, 226–793 (1960).
24 WHITAKER, S., The transport equations for multiphase systems, Chem. Engng. Sci., Vol. 28, 139–147 (1973).
25 ZUBER, N., On the dispersed two-phase flow in a laminar flow region, Chem. Engng. Sci., Vol. 19, 897–917 (1964).

Eingereicht am 8. 2. 1979

Anschrift: Dr.-Ing. HANS IMMICH, BBC Aktiengesellschaft BROWN, BOVERI & Cie., Abteilung TGD-14, CH-5401 Baden, Schweiz

KLEINE MITTEILUNGEN

ZAMM 60, 109 – 111 (1980)

T. Vajda / E. V. Nagy

Ein Algorithmus zur approximativen Lösung von Systemen gewöhnlicher, homogener, nichtlinearer Differentialgleichungen zweiter Ordnung in der chemischen Verfahrenstechnik

Die Arbeit beschreibt eine heuristische Methode zur analytischen Approximation des Diffusionsmodells eindimensionaler, nichtlinearer chemischer Transportsysteme mit axialer Vermischung. Es wird mittels Fehlerabschätzung gezeigt, daß die Methode in für die Praxis wichtigen Fällen hinreichend genau ist.

I. Einleitung

In der Verfahrenstechnik tritt häufig das Problem der mathematischen Beschreibung solcher eindimensionaler, stationärer Mehrphasentransportsysteme mit axialer Mischung auf, die durch ein nur vom Wert der betreffenden Potentiale abhängendes Quellenglied von beliebiger Form charakterisiert werden können. Bekanntlich erfolgt diese mathematische Beschreibung mit Hilfe gewöhnlicher, homogener, nichtlinearer Differentialgleichungssysteme zweiter Ordnung mit den zugehörigen, teils inhomogenen Randbedingungen, falls die Mischung durch den Formalismus der sogenannten Diffusionsmodelle berücksichtigt wird. Vom Standpunkt der Verfahrenstechnik aus, die es vorzieht, mit Gleichungen erster Ordnung zu arbeiten, ergeben sich durch die Systeme zweiter Ordnung im Problem jedoch gewisse Schwierigkeiten in der Interpretation, deren Lösung zweckmäßig zu sein scheint. Außerdem ist die Möglichkeit zur Gewinnung sonst vergleichbarer analytischer Lösungen für Probleme zweiter Ordnung viel beschränkter als für solche erster Ordnung.

Obwohl sich bereits viele Autoren (vgl. [1–6]) mit der approximativen Lösung mathematischer Modelle für chemische Transportprozesse verschiedener Typen beschäftigt haben, wurde von den bisher durchgeführten Untersuchungen obiger Fragenkreis nicht berührt. In unserer Arbeit stellen wir durch eine entsprechende heuristische Umdeutung der Aufgabe zweiter Ordnung, die der mathematischen Formulierung der Diffusionsmodelle dient, ein approximatives mathematisches Modell auf, das in diesem Sinne eine analytische Beschreibung des untersuchten Transportsystems darstellt und aus Gleichungen erster Ordnung besteht. Indem man aus beiden Aufgaben Integralgleichungssysteme bildet und miteinander vergleicht, kann man erkennen, daß die durch das Gleichungssystem erster Ordnung erzeugten Lösungen tatsächlich als approximative Lösungen des Gleichungssystems zweiter Ordnung interpretiert werden können.

II. Die Grundaufgabe und ihre Approximation

Aus der Literatur (z. B. [7]) ist bekannt, daß Transportprozesse der oben angegebenen Systeme gewöhnlich durch das auf Normalform transformierte Gleichungssystem

$$-A_i \frac{d^2u_i}{dx^2} + \frac{du_i}{dx} - f_i(u_1, \dots, u_n) = 0 \tag{1}$$

mit den Randbedingungen

$$u_i(0) - A_i \left(\frac{du_i}{dx}\right)_{x=0} = u_{ei}\,, \tag{2}$$

und

$$\left(\frac{du_i}{dx}\right)_{x=1} = 0\,, \qquad i = 1, 2, \dots, n\,, \tag{3}$$

beschrieben werden können (der Einfachheit halber betrachten wir hier nur Gleichstromsysteme, es ist aber leicht zu sehen, daß sich die folgenden Ableitungen auch für Gegenstromsysteme herleiten ließen). In (1)–(3) sind A_i, $0 < A_i < +\infty$, die verallgemeinerten Mischungsfaktoren, x, $0 \leqq x \leqq 1$ ist die Normalkoordinate; die u_i bezeichnen die gesuchten Potentialfunktionen, die zu den n extensiven Mengen gehören, deren Transport untersucht werden soll, während u_{ei} die Eingangspotentialwerte sind. Für die Quellenfunktionen f_i nehmen wir während des gegebenen Prozesses Beschränktheit über der Menge der betrachteten Potentialwerte an. Sie mögen überdies in allen Variablen einer Lipschitz-Bedingung genügen. Dann ist offenbar Existenz und Eindeutigkeit der durch (1)–(3) erzeugten Potentialfunktionen gesichert.

Wir betrachten nun das auf Grund heuristischer Überlegungen konstruierte Differentialgleichungssystem

$$(1 + A_i)\,\frac{dv_i}{dx} - f_i(v_1, \dots, v_n) = 0 \tag{4}$$

mit den zugehörigen Randbedingungen

$$(1 + A_i)\, v_i(0) - A_i v_i(1) = u_{ei}\,, \qquad i = 1, 2, \dots, n\,, \tag{5}$$

in dem die Funktionen v_i der Aufgabe (4)–(5) genügen, während die restlichen Bezeichnungen unverändert bleiben. Ziel unserer weiteren Untersuchungen ist eine Klärung der Frage, wieweit und mit welcher Genauigkeit die v_i approximativen Lösungen von Aufgabe (1)–(3) entsprechen.

Unter diesem Gesichtspunkt führt ein direkter Vergleich der Aufgaben (1)–(3) und (4), (5) zu keinem Resultat. Es ist aber möglich, durch formales, implizites Aufschreiben der Lösungen Integralzusammenhänge zu gewinnen, die als Integralgleichungsinverse der linearen Anteile der Differentialgleichungen angesehen werden können. Sie enthalten auch die Randbedingungen und sind dadurch mit der Grundaufgabe bzw. deren Approximation äquivalent. Stellt man die Basissysteme der in den Gleichungen (1) und (4) auftretenden Differentialoperatoren fest und wendet man die Methode der Variation der Konstanten ([8]) an, so erhält man die Integralgleichungen für $i = 1, 2, \dots, n$

$$u_i(x) = u_{ei} + \int_0^x f_i(u_1(\xi), \dots, u_n(\xi))\, d\xi + \int_x^1 e^{\frac{x-\xi}{A_i}} f_i(u_1(\xi), \dots, u_n(\xi))\, d\xi\,, \tag{6}$$

$$v_i(x) = u_{ei} + \int_0^x f_i(v_1(\xi), \dots, v_n(\xi))\, d\xi + \frac{A_i}{1 + A_i} \int_x^1 f_i(v_1(\xi), \dots, v_n(\xi))\, d\xi\,. \tag{7}$$

Die formale Ähnlichkeit von (6) und (7) liegt auf der Hand, und diese Tatsache hilft erklären, was wir unter *heuristischen Überlegungen* verstehen. Zuerst wurde nämlich die Aufgabe (1)–(3) mit der obengenannten Methode in Integralgleichung (6) umgeformt, und es wurde untersucht, ob es möglich ist, eine solche Approximation von $e^{\frac{x-\xi}{A_i}}$ $(i = 1, \dots, n)$ zu finden, die sich als Funktion der Variablen A_i ähnlich verhält, und für die die mit dieser approximativen Form gebildete Integralgleichung entsprechende Randwertaufgabe möglichst einfach, d. h. von erster Ordnung ist. Unter Berücksichtigung obiger Gesichtspunkte wählten wir die Funktion $\frac{A_i}{1 + A_i}$ $(i = 1, \dots, n)$ als Approximation. Es ist einfach zu beweisen, daß der Wert $\int_0^1 \int_x^1 \left| e^{\frac{x-\xi}{A_i}} - \frac{A_i}{1 + A_i} \right| d\xi\, dx$ für jeden Wert von A_i klein genug ist und dadurch im wesentlichen garantiert wird, daß die Aufgabe (4), (5) eine gute Approximation der Aufgabe (1)–(3) darstellt. Es wäre interessant zu untersuchen, ob man eine solche Funktion finden könnte, die $e^{\frac{x-\xi}{A}}$ besser approximiert und für die gleichzeitig die entsprechende Randwertaufgabe von erster Ordnung bleibt. Angesichts der Notwendigkeit, daß beide Bedingungen erfüllt sein müssen, weist diese Frage ernsthafte Schwierigkeiten auf. Dies überschreitet aber unsere Zielsetzung.

Es sei schließlich bemerkt, daß die von den bei Gleichungen erster Ordnung üblichen Anfangsbedingungen abweichende Form der Bedingung (5) ein wichtiges Moment der heuristischen

Approximation ist. Man sieht jedoch ein, daß für die vorgegebenen Quellenfunktionen f_i bei den für die Praxis wichtigen Fällen die aus (4), (5) bestimmbaren approximativen Potentialfunktionen existieren und eindeutig sind.

III. Abschätzung des Fehlers der Approximation

Die Genauigkeit der Approximationsmethode untersuchen wir durch Abschätzung der Größen

$$\delta_i(x) = |u_i(x) - v_i(x)|\,, \qquad i = 1, 2, \ldots, n\,. \tag{8}$$

Nach Einsetzen von (6) und (7) in (8), Vertauschen von Absolutstrichen und Integrationszeichen sowie Erweitern ergibt sich

$$\begin{aligned}\delta_i(x) \leqq \int_0^x |f_i(u_1, \ldots, u_n) - f_i(v_1, \ldots, v_n)|\, d\xi \,+ \\ + \frac{A_i}{1 + A_i}\int_x^1 |f_i(u_1, \ldots, u_n) - f_i(v_1, \ldots, v_n)|\, d\xi \,+ \\ + \int_x^1 \left|\left(e^{\frac{x-\xi}{A_i}} - \frac{A_i}{1 + A_i}\right) f_i(u_1, \ldots, u_n)\right| d\xi\,, \\ i = 1, 2, \ldots, n\,.\end{aligned} \tag{9}$$

Nach Voraussetzung existiert:

$$M_i = \max\,[|f_i(u_1, \ldots, u_n)|]\,, \tag{10}$$

und es gilt für $i = 1, 2, \ldots, n$

$$|f_i(u_1, \ldots, u_n) - f_i(v_1, \ldots, v_n)| \leqq K_i \sum_{j=1}^n |u_j - v_j|\,, \tag{11}$$

wobei die Größen K_i und M_i für die konkreten f_i geeignete Konstanten sind, die sich angeben lassen. Mit Hilfe von (8), (10) und (11) läßt sich (9) weiter umformen zu

$$\begin{aligned}\delta_i(x) \leqq K_i \int_0^x \sum_{j=1}^n \delta_j(\xi)\, d\xi + \frac{K_i A_i}{1 + A_i}\int_x^1 \sum_{j=1}^n \delta_j(\xi)\, d\xi \,+ \\ + M_i \int_x^1 \left| e^{\frac{x-\xi}{A_i}} - \frac{A_i}{1 + A_i}\right| d\xi\,, \quad i = 1, 2, \ldots, n\,.\end{aligned} \tag{12}$$

Integriert man die dritten Glieder im System (12), so folgt

$$\varphi_i(x, A_i, M_i) = M_i \int_x^1 \left| e^{\frac{x-\xi}{A_i}} - \frac{A_i}{1 + A_i}\right| d\xi =$$

$$= \begin{cases} M_i A_i\left(1 + e^{\frac{x-1}{A_i}}\right) + \dfrac{M_i A_i}{1 + A_i}(1 - x) + \dfrac{2M_i A_i^2}{1 + A_i}\left(\ln \dfrac{A_i}{1 + A_i} - 1\right), \\ \qquad \text{falls } 0 \leqq x \leqq 1 + A_i \ln \dfrac{A_i}{1 + A_i}\,, \\ M_i A_i\left(1 - e^{\frac{x-1}{A_i}}\right) + \dfrac{M_i A_i}{1 + A_i}(x - 1)\,, \\ \qquad \text{falls } 1 + A_i \ln \dfrac{A_i}{1 + A_i} < x \leqq 1\,, \end{cases} \tag{13}$$

für $i = 1, 2, \ldots, n$. Man sieht leicht, daß die Funktionen φ_i für $A_i \to 0$ bzw. $A_i \to +\infty$ die folgende sehr wichtige Eigenschaft haben:

$$\lim_{A_i \to 0} \varphi_i(x, A_i, M_i) = \lim_{A_i \to +\infty} \varphi_i(x, A_i, M_i) = 0 \tag{14}$$

für $i = 1, 2, \ldots, n$.

Wir führen nun den *integrierten Gesamtfehler*

$$\Delta = \Delta(x) = \int_0^x \sum_{i=1}^n \delta_i(\xi)\, d\xi\,, \tag{15}$$

sowie die Zahlen

$$\alpha = \sum_{i=1}^n \frac{K_i}{1 + A_i}\,, \quad \beta = \sum_{i=1}^n \frac{A_i K_i}{1 + A_i}\,, \quad \Phi(x) = \sum_{i=1}^n \varphi_i(x) \tag{16}$$

ein. Dann ergibt sich nach Addition der Ungleichungen des Systems (12) unter Berücksichtigung von (15) und (16) die folgende Differentialungleichung:

$$\frac{d\Delta}{dx} \leqq \alpha\Delta + \beta\Delta(1) + \Phi \tag{17}$$

mit der Anfangsbedingung $\Delta(0) = 0$. Integration der Ungleichung (17) von 0 bis x liefert die Relation

$$\Delta(x) \leqq \alpha \int_0^x \Delta(\xi)\, d\xi + \beta x \Delta(1) + \int_0^x \Phi(\xi)\, d\xi\,. \tag{18}$$

Wird jetzt (18) mit Hilfe des GRONWALLschen Lemmas (vgl. [10]) gelöst, so ergibt sich die Abschätzung

$$\Delta(x) \leqq \frac{\beta}{\alpha}\Delta(1)\,(e^{\alpha x} - 1) + \int_0^x e^{\alpha(x-\xi)}\,\Phi(\xi)\, d\xi\,. \tag{19}$$

Im Fall $x = 1$ gewinnen wir aus (19)

$$\Delta(1)\,[\alpha - \beta(e^\alpha - 1)] \leqq \alpha \int_0^1 e^{\alpha(1-\xi)}\,\Phi(\xi)\, d\xi\,.$$

Ist die Ungleichung $\beta(e^\alpha - 1) < \alpha$ erfüllt, so kann man hieraus, wiederum für $x = 1$, eine Abschätzung für $\Delta(1)$ erhalten:

$$\Delta(1) \leqq \frac{\alpha}{\alpha - \beta\,(e^\alpha - 1)} \int_0^1 e^{\alpha(1-\xi)}\,\Phi(\xi)\, d\xi\,. \tag{20}$$

Betrachten wir noch die Bedingung $\beta(e^\alpha - 1) < \alpha$. Sie bedeutet für die verallgemeinerten Mischungsfaktoren A_i, $i = 1, 2, \ldots, n$, und für die den gegebenen Quellenfunktionen entsprechenden Konstanten K_i, $i = 1, 2, \ldots, n$,

$$\sum_{i=1}^n K_i > e^{\sum_{i=1}^n \frac{K_i}{1 + A_i}} \sum_{i=1}^n \frac{A_i K_i}{1 + A_i}\,.$$

Diese Bedingung ist für hinreichend kleine A_i stets erfüllt, bedeutet aber eine Einschränkung für die K_i und damit für die Funktionen f_i.

Wenn wir die Abschätzung (20) in die rechte Seite von (19) einsetzen, dann erhalten wir

$$\Delta(x) \leqq \frac{\beta(e^{\alpha x} - 1)}{\alpha - \beta(e^\alpha - 1)} \int_0^1 e^{\alpha(1-\xi)}\,\Phi(\xi)\, d\xi + \int_0^x e^{\alpha(x-\xi)}\,\Phi(\xi)\, d\xi\,. \tag{21}$$

Durch Einsetzen der Ergebnisse (20) und (21) in (17) kommen wir schließlich zu der gewünschten Abschätzung für $d\Delta/dx$:

$$\begin{aligned}\sum_{i=1}^n \delta_i(x) \equiv \frac{d\Delta}{dx} \leqq \frac{\beta\alpha\, e^{\alpha x}}{\alpha - \beta(e^\alpha - 1)} \int_0^1 e^{\alpha(1-\xi)}\,\Phi(\xi)\, d\xi \,+ \\ + \alpha \int_0^x e^{\alpha(x-\xi)}\,\Phi(\xi)\, d\xi + \Phi(x)\,.\end{aligned} \tag{22}$$

Aus der Beziehung (22) kann geschlossen werden, daß wegen der Eigenschaften, die in (14) angegeben wurden, der Fehler der Approximationsmethode, d. h. also der Unterschied zwischen den Ergebnissen, die sich aus den Aufgaben erster bzw. zweiter Ordnung ableiten, gegen Null konvergiert, falls die in einem beliebigen System gewählten Mischungsfaktoren A_i gegen Null oder gegen Unendlich streben, sofern im letzten Fall $\alpha + \beta \leqq 1$ ist. Dies sind die in der Praxis sehr wichtigen Fälle, in denen überhaupt keine oder eine sehr vollständige Mischung auftritt. Wie man auch aus der Abbildung 1 ersehen kann, läßt die numerische Analyse der in (13) eingeführten Funktionen φ_i die Vermutung zu, daß der Fehler im allgemeinen relativ gering ist, wenn sich nicht allzu schnell ändernde Quellenfunktionen f_i von niedrigem Wert vorliegen.

Zusammenfassend stellen wir fest, daß die hier vorgestellten Resultate für die rechentechnische Praxis und Ingenieurarbeit eine neue Interpretation der üblichen Diffusionsmodelle liefern. Andererseits zeigt die Anwendung der Gleichungen erster und zweiter Ordnung in hier nicht aufgeführten numerischen Bei-

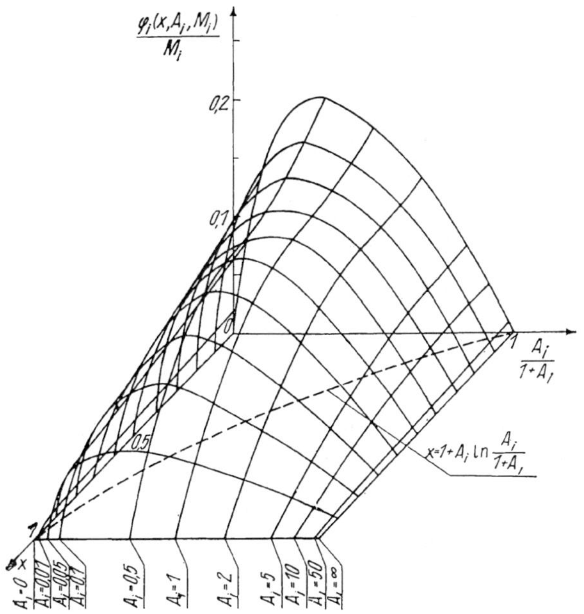

Abb 1

spielen mit gleicher Zielsetzung und der Vergleich der Resultate, daß die Methode in den für die Praxis wichtigen Parameterintervallen viel genauer ist, als auf Grund der Fehlerabschätzung zu erwarten wäre (vgl. [9]).

Die Verfasser möchten auf diese Weise ihren Dank für die Hilfe des Gutachters bei der entgültigen Ausgestaltung der Arbeit ausdrücken.

Literatur

1 KONDRATIEV, G M (Кондратьев, Г. М.) Регулярный тепловый режим, Москва 1954.
2 HEILIGENSTÄDT, W , Wärmetechnische Rechnungen für Industrieöfen, Stahleisen, Düsseldorf 1951
3 SASVÁRI, G , Acta Chimica Acad Sci Hungar 40, 343 (1964)
4 POLINSZKY, K., et al , Acta Chimica Acad Sci. Hungar 85, 161 (1975).
5 – , Hungar J Indust. Chem. 4, 151 (1976).
6 FÜLÖP, J , et al , Preprint Ин-т тепло- и массооб'мена, Акад. Наук БССР, Минск 1977
7 BIRD, R. B , et al , Transport Phenomena, Wiley, New York 1966
8 KNESCHKE, A , Differentialgleichungen I , Teubner, Leipzig 1965
9 VAJDA, T , Kand Diss. (eingereicht)
10 WALTER, W , Differential and Integral Inequalities, Springer, Berlin, Heidelberg, New York 1970.

Eingereicht: 20. 4. 1978, endgültige Fassung: 19. 2. 1979

Anschriften: Dr. TÁDÉ VAJDA, Forschungsinstitut für Technische Chemie der Ungarischen Akademie der Wissenschaften, H-1502 Budapest, P.O.B. 98, Ungarn;

Dr. ÉVA V. NAGY, Technische Universität Budapest, Lehrstuhl für Mathematik der Fakultät für Elektrizität, H-1111 Budapest, Stoczek u. 4., H. ép. III em., Ungarn

ZAMM 60, 111 – 112 (1980)

G. MERZ

Eine Kettenbruchdarstellung für die Norm des kubischen periodischen Spline-Interpolationsoperators

1. Es sei C_N der lineare Raum der auf der reellen Achse stetigen Funktionen mit der Periode $N \in \mathbb{N}$ und S_N^3 der lineare Raum der kubischen N-periodischen Spline-Funktionen, deren Knoten in den ganzen Zahlen liegen. Maßgebend für die Beurteilung der numerischen Brauchbarkeit des durch den Spline-Interpolationsoperator $\mathcal{L}_N^3(f)$: $C_N \to S_N^3$ erzeugten Approximationsprozesses ist die Größe der durch

$$\|\mathcal{L}_N^3\| = \sup_{\substack{\|f\| \leqq 1 \\ f \in C_N}} \|\mathcal{L}_N^3(f)\|$$

festgelegten Norm von $\mathcal{L}_N^3$. Dies folgt unmittelbar aus der für jedes $f \in C_N$ gültigen Abschätzung

$$\|\mathcal{L}_N^3(f) - f\| \leqq (1 + \|\mathcal{L}_N^3\|) \inf_{s \in S_N^3} \|f - s\| .$$

Im folgenden soll für den Fall der L_1-Norm eine sehr einfache Möglichkeit zur rekursiven Berechnung von $\|\mathcal{L}_N^3\|_1$ für beliebiges ungerades N aufgezeigt werden. Den Ausgangspunkt unserer Überlegungen bildet dabei die in [1] angegebene Darstellung von $\|\mathcal{L}_N^3\|_1$ als komplexes Linienintegral. Mit dem verallgemeinerten EULER-FROBENIUS-Polynom

$$H_3(t, z) = t^3(1 - z)^3 + 3t^2z(1 - z)^2 + 3tz(1 - z^2) + z(z^2 + 4z + 1) ,$$

der Matrix der diskreten FOURIER-Transformation

$$W = \frac{1}{\sqrt{N}} ((\zeta^{\mu\nu}))_0^{N-1} , \qquad \zeta = \exp \frac{2\pi i}{N} ,$$

und der Diagonalmatrix

$$Q(t) = \operatorname{diag} \frac{H_3(t, \zeta^\mu)}{H_3(1, \zeta^\mu)} , \qquad \mu = 0(1) N - 1 ,$$

gilt für den zu $f \in C_N$ gehörigen Interpolationsspline $s \in S_N^3$

$$s(t) = W^* Q(t) W y , \qquad y = \begin{pmatrix} f(0) \\ \vdots \\ f(N-1) \end{pmatrix},$$

und hieraus folgt fur $N = 2n + 1$ durch Residuenrechnung

$$\|\mathcal{L}_{2n+1}^3\|_1 = \frac{1}{4\pi i} \oint_C \frac{1 + z^{2n+1}}{1 - z^{2n+1}} \frac{z^2 + 10z + 1}{zH_3(1, z)} dz =$$

$$= \frac{1}{2} \left[1 + \sqrt{3} \frac{1 - (2 - \sqrt{3})^{2n+1}}{1 + (2 - \sqrt{3})^{2n+1}} \right] . \tag{1}$$

C bezeichnet dabei den positiv durchlaufenen Kreis $|z| = \varrho$, $\varrho < 1$, der die Nullstelle $z_1 = -2 + \sqrt{3}$ von $H_3(1, z) = z^2 + 4z + 1$ umschließt. Aus (1) entnimmt man unmittelbar die Monotonieaussage

$$\|\mathcal{L}_1^3\|_1 < \|\mathcal{L}_3^3\|_1 < \|\mathcal{L}_5^3\|_1 < \|\mathcal{L}_7^3\|_1 < \dots < \tfrac{1}{2} (1 + \sqrt{3}) . \tag{2}$$

Eine einfache Umformung von (1) führt auf

$$\|\mathcal{L}_{2n+1}^3\|_1 = \frac{A_n}{B_n}$$

mit

$$A_n = \frac{\sqrt{3} + 2}{2\sqrt{3}} (2 + \sqrt{3})^n + \frac{\sqrt{3} - 2}{2\sqrt{3}} (2 - \sqrt{3})^n ,$$

$$B_n = \frac{\sqrt{3} + 1}{2\sqrt{3}} (2 + \sqrt{3})^n + \frac{\sqrt{3} - 1}{2\sqrt{3}} (2 - \sqrt{3})^n .$$

Offenbar genügen A_n und B_n für $n \geqq 0$ den Differenzengleichungen

$$A_{n+2} - 4A_{n+1} + A_n = 0 \tag{3}$$

und

$$B_{n+2} - 4B_{n+1} + B_n = 0 \tag{4}$$

mit den Anfangsbedingungen $A_0 = 1$, $A_1 = 4$, $B_0 = 1$, $B_1 = 3$. Man beachte, daß die charakteristische Gleichung von (3) bzw. (4) gerade $H_3(1, -z) = 0$ ist.

Aus (3) und (4) folgt ferner $A_{n+1}B_n - B_{n+1}A_n = 1$, was analog wie bei FAREY-Folgen

$$\frac{A_n}{B_n} = \frac{A_{n-1} + A_{n+1}}{B_{n-1} + B_{n+1}}$$

nach sich zieht:

$\frac{A_n}{B_n}$ ist Mediante von $\frac{A_{n-1}}{B_{n-1}}$ und $\frac{A_{n+1}}{B_{n+1}}$.

n	$\|\|\mathcal{L}^3_{2n+1}\|\|_1$	Numer. Wert
0	1	1.00000000
1	$\frac{4}{3}$	1.33333333
2	$\frac{15}{11}$	1.36363636
3	$\frac{56}{41}$	1.36585365
4	$\frac{209}{153}$	1.36601307
5	$\frac{780}{571}$	1.36602451
6	$\frac{2911}{2131}$	1.36602534
7	$\frac{10\,864}{7\,953}$	1.36602539
8	$\frac{40\,545}{29\,681}$	1.36602540
	...	...
	$\frac{1}{2}(1+\sqrt{3})$	1.36602540

2. Die in (2) angegebene obere Schranke besitzt die Kettenbruchdarstellung

$$\frac{1}{2}(1+\sqrt{3}) = 1 + \frac{1}{2+}\,\frac{1}{1+}\,\frac{1}{2+}\,\frac{1}{1+}\,\frac{1}{2+}\cdots = [\overline{1,2}]\,. \tag{5}$$

Satz 1: *Bezeichnet* $\frac{p_n}{q_n}$ *den Näherungsbruch n-ter Ordnung des Kettenbruchs* (5), *so gilt für* $n = 0, 1, 2, \ldots$

$$\|\mathcal{L}^3_{2n+1}\|_1 = \frac{p_{2n}}{q_{2n}}\,.$$

Beweis: Die Näherungsbrüche von (5) berechnen sich für $\nu \geqq 1$ aus den Anfangswerten $p_0 = 1$, $q_0 = 1$, $p_1 = 3$, $q_1 = 2$ vermöge

$$p_{2\nu} = p_{2\nu-1} + p_{2\nu-2} \qquad \text{bzw.} \qquad p_{2\nu+1} = 2p_{2\nu} + p_{2\nu-1}$$
$$q_{2\nu} = q_{2\nu-1} + q_{2\nu-2} \qquad\qquad q_{2\nu+1} = 2q_{2\nu} + q_{2\nu-1}\,.$$

Hieraus folgt für $\nu \geqq 2$

$$p_{2\nu} = (2p_{2\nu-2} + p_{2\nu-3}) + p_{2\nu-2} = 4p_{2\nu-2} - (p_{2\nu-2} - p_{2\nu-3}) = $$
$$= 4p_{2\nu-2} - p_{2\nu-4}\,.$$

Eine analoge Beziehung besteht für die Näherungsnenner. Der Vergleich mit (3) und (4) ergibt wegen $p_2 = 4$, $q_2 = 3$ die Behauptung.

3. Die Zahl $\frac{1}{2}(\sqrt{3} + 1)$ ist Nullstelle des Polynoms

$$P(x) = 2x^2 - 2x - 1\,. \tag{6}$$

Wir zeigen, daß die Normen $\|\mathcal{L}^3_{2n+1}\|_1$ als sukzessive Näherungen eines geeigneten Iterationsverfahrens zur Bestimmung dieser Nullstelle aufgefaßt werden können.

Dazu beachten wir, daß die Regula Falsi zur iterativen Lösung von (6) für die feste Ausgangsnäherung $x_{-1} = \frac{3}{2}$ und eine beliebige zweite Ausgangsnäherung x_0 mit $P(x_0) < 0$ die Form

$$x_1 = \frac{3x_0 + 1}{2x_0 + 1}$$

annimmt.

Satz 2: *Startet man die Regula Falsi*

$$x_{n+1} = \frac{3x_n + 1}{2x_n + 1} \tag{7}$$

mit $x_0 = 1$, *so gilt für* $n = 0, 1, 2, \ldots$

$$\|\mathcal{L}^3_{2n+1}\|_1 = x_n\,.$$

Beweis: Setzt man in (7) $x_n = \frac{r_n}{s_n}$ mit $r_n, s_n \in \mathbb{N}$ und $(r_n, s_n) = 1$, so folgt

$$r_{n+1} = 3r_n + s_n\,, \qquad s_{n+1} = 2r_n + s_n\,,$$

und hieraus ergibt sich für $n \geqq 0$

$$r_{n+2} = 4r_{n+1} + (s_{n+1} - r_{n+1}) = 4r_{n+1} - r_n$$

sowie

$$s_{n+2} = 4s_{n+1} + (2r_{n+1} - 3s_{n+1}) = 4s_{n+1} - s_n\,.$$

Wegen $r_0 = s_0 = 1$, $r_1 = 4$, $s_1 = 3$ und (3), (4) ist damit schon alles gezeigt.

4. Über entsprechende Aussagen bei Splines höheren Grades sowie für andere Normen soll in einer weiteren Veröffentlichung berichtet werden. Beispielsweise hängt die TSCHEBYSCHEFF-Norm von $\mathcal{L}^3_N$ mit den Näherungsbrüchen des Kettenbruches

$$[1, \overline{1, 1, 4, 1, 1, 2, 20, 2}]$$

zusammen.

Literatur

1 MEINARDUS, G.; MERZ, G., Zur periodischen Spline-Interpolation II. Erschienen in: L. COLLATZ / G. MEINARDUS / H. WERNER (eds.): Numerische Methoden der Approximationstheorie Band 4 (ISNM 42), Basel und Stuttgart: Birkhäuser Verlag 1978, S. 204–221.

Eingereicht: 6. 11. 1978, korrigierte Fassung 9. 4. 1979

Anschrift: Prof. Dr. GERHARD MERZ, Gesamthochschule Kassel, Fachbereich 17 — Mathematik, Wilhelmshöher Allee 73, D-3500 Kassel, BRD

ZAMM 60, 112–113 (1980)

R. N. KAUL / B. GUPTA

Efficiency and Linear Vector Maximum Value Problem

1. Introduction

A vector maximization problem arises when more than one scalar valued objective function is to be maximized over a given feasibility region. The concepts of efficiency and proper efficiency [2, 4, 5] play a useful role in the analysis of such problems. This note considers a linear fractional vector maximum value problem and derives a necessary condition for a point to be efficient which, however, is not sufficient. For this a result by FOCKE [1] is made use of. The corresponding condition for the linear vector maximum problem is both necessary and sufficient [3].

2. Linear fractional vector maximum problem

Consider the linear fractional vector maximization problem, namely

$$\text{(LFMP)} \quad h(x) = \left(\frac{c_1^{\mathsf T}x}{d_1^{\mathsf T}x}, \frac{c_2^{\mathsf T}x}{d_2^{\mathsf T}x}, \ldots, \frac{c_k^{\mathsf T}x}{d_k^{\mathsf T}x}\right) = \max!$$

subject to $x \in X$, where

$$X = \{x \in R^n\colon Ax \leqq b,\ x \geqq 0\}\,.$$

Here A is an $m \times n$ matrix, b is an $m \times 1$ vector; also the c_i's and the d_i's are $n \times 1$ vectors for all $i = 1, \ldots, k$. It is also assumed that the functions $d_i^{\mathsf T}x$ (for all $i = 1, \ldots, k$) are all positive over X.

We have the following definition for efficiency.

Definition: A point x^0 is said to be *efficient* for the problem (LFMP) above, if $x^0 \in X$ and there exists no other point $x^* \in X$ such that $h(x^*) \geqq h(x^0)$ and $h(x^*) \neq h(x^0)$.

To the above (LFMP), we associate the following linear vector maximum problem,

$$l(x) = (c_1^{\mathsf T}x, c_2^{\mathsf T}x, \ldots, c_k^{\mathsf T}x, -d_1^{\mathsf T}x, -d_2^{\mathsf T}x, \ldots, -d_k^{\mathsf T}x) = \text{Max!}$$

(LVMP)
subject to $x \in X$.

We now have the following result:

Theorem: *If x^0 is an efficient solution of problem* (LFMP) *then it is efficient for problem* (LVMP).

Proof: Let x^0 be an efficient solution of (LFMP). Suppose that x^0 were not an efficient solution of (LVMP). Then there would exist an $x^* \in X$ and a certain $j \in \{1, 2, \ldots, k\}$, such that $c_i^{\mathsf{T}} x^* \geqq c_i^{\mathsf{T}} x^0$ and $d_i^{\mathsf{T}} x^* \leqq d_i^{\mathsf{T}} x^0$ for all i and $c_j^{\mathsf{T}} x^* > c_j^{\mathsf{T}} x^0$ or $d_j^{\mathsf{T}} x^* < d_j^{\mathsf{T}} x^0$ which would imply that

$$\frac{c_i^{\mathsf{T}} x^*}{d_i^{\mathsf{T}} x^*} \geqq \frac{c_i^{\mathsf{T}} x^0}{d_i^{\mathsf{T}} x^0} \quad \text{for all } i \text{ and } \quad \frac{c_j^{\mathsf{T}} x^*}{d_j^{\mathsf{T}} x^*} > \frac{c_j^{\mathsf{T}} x^0}{d_j^{\mathsf{T}} x^0},$$

contrary to the efficiency of x^0, with respect to (LFMP). Hence the theorem.

The following example shows that the converse of the above theorem does not necessarily hold.

Example: Assume $n = 2$ and $k = 1$.

Let $h(x) = \dfrac{x_1}{x_2} = \max!$, subject to $x \in X$

where X is the feasible region, determined by the linear inequalities:

$$x_1 + x_2 \geqq 4\,,$$
$$-x_1 + 3x_2 \geqq 3\,,$$
$$2x_1 - x_2 \leqq 9\,,$$
$$-2x_1 + 5x_2 \leqq 19\,, \quad \text{as shown in the figure 1}.$$

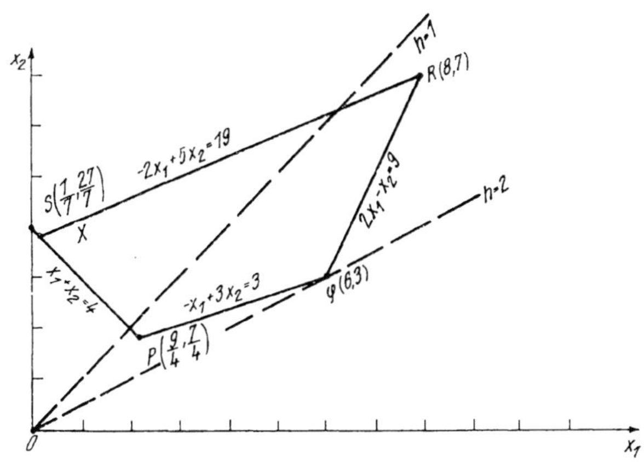

Figure 1

It can be easily verified that the optimal value of $h(x)$ is 2, at the point $Q(6,3)$.

Let (x_1^0, x_2^0) be a point on the edge PQ of the set X i.e., $-x_1^0 + 3x_2^0 = 3$. We assert that it is efficient for (LVMP). Assume the existence of $(x_1^*, x_2^*) \in X$, such that either $x_1^* \geqq x_1^0$ and $x_2^* < x_2^0$ or $x_1^* > x_1^0$ and $x_2^* \leqq x_2^0$. In either case, we obtain, $-x_1^* + 3x_2^* < 3$, which shows that the second inequality above, determining X, is violated, a contradiction to our assumption, that $(x_1^*, x_2^*) \in X$. Hence every point on the edge PQ, joining P to Q, of X, is efficient.

Similarly, we can show that every point on the edge of X, joining Q to R, is also efficient.

Corollary to the Theorem: *If x^0 is an efficient solution of* (LFMP), *then it is necessary that there exist $u^0 > 0$ and $v^0 > 0$, such that*

$$\sum_{i=1}^{k} (u_i^{0\mathsf{T}} c_i^{\mathsf{T}} - v_i^0 d_i^{\mathsf{T}})\, x^0 \geqq \sum_{i=1}^{k} (u_i^0 c_i^{\mathsf{T}} - v_i^0 d_i^{\mathsf{T}})\, x \quad \textit{for all} \quad x \in X\,.$$

Proof: As x^0 is efficient for (LFMP), it is also efficient for (LVMP). Following Geoffrion [2] and Focke [1], therefore x^0 is optimal for the linear parametric programming problem,

$$\sum_{i=1}^{k} (u_i c_i^{\mathsf{T}} - v_i d_i^{\mathsf{T}})\, x = \text{Max!}\,, \quad \text{subject to } x \in X\,,$$

for certain parameters $u^0 > 0$ and $v^0 > 0$. Thus,

$$\sum_{i=1}^{k} (u_i^0 c_i^{\mathsf{T}} - v_i^0 d_i^{\mathsf{T}})\, x^0 \geqq \sum_{i=1}^{k} (u_i^0 c_i^{\mathsf{T}} - v_i^0 d_i^{\mathsf{T}})\, x \quad \text{for all } x \in X\,,$$

which is the required necessary condition, for a point to be efficient for (LFMP).

Acknowledgement

The authors are grateful to the referee for useful suggestions and bringing this reference [1] to their attention.

References

1 Focke, J., Vectormaximumproblem und parametrische Optimierung, Math. Operationsforsch. Statist. 4, 365–369 (1973).
2 Geoffrion, A. M., Proper efficiency and the theory of vector maximization, J. Math. Analysis Appl. 22, 618–630 (1968).
3 Isermann, H., Proper efficiency and the linear vector maximum problem, Operations Res. 22, 189–191 (1974).
4 Klinger, A., Improper solutions of the vector maximum problem, Operations Res. 15, 570–572 (1967).
5 Kuhn, H. W.; Tucker, A. W., Nonlinear programming, in J. Neymann (ed.) Proc. Second Berkeley Sympos. on Math. Statist. and Probab., Univ. of California Press, Berkeley, Cal. 1951, pp. 481–492.

Eingereicht: 25. 6. 1977, endgültige Fassung: 23. 3. 1979

Anschriften: Prof. Dr. R. N. Kaul, Head of the Department of Mathematics, University of Delhi, Delhi-110007, India;
Mrs. Bina Gupta, Department of Mathematics, University of Delhi, Delhi-110007, India.

ZAMM 60, 113–114 (1980)

M. C. Nucci

Rotation Motions of a Gyrostatic Satellite in a Kepler Orbit

In the present paper, the rotation motions about the mass center are studied for a „Volterra-type" gyrostatic satellite in a Kepler orbit under a Newtonian force field. All the possible stationary rotations are determined. Moreover it is found a class of non-stationary rotations.

1. Equations of motion

Denote by S a "Volterra-type" gyrostatic satellite [1] with gyrostatic momentum $\boldsymbol{I}$, suppose that its mass center G describes a Kepler orbit under a Newtonian force field (2nd order approximation [2]) with respect to the fixed center O and let $\boldsymbol{e}$ be the unit vector of the normal to the orbit plane: $\mathrm{d}OG/\mathrm{d}t = \nu \boldsymbol{e} \wedge OG$ with ν a right scalar function of the time t. The motion around the mass center is supposed to be a rotation; in other words the angular velocity $\boldsymbol{\omega}$ with respect to the mass center is given by:

$$\boldsymbol{\omega} = \omega \boldsymbol{\chi}$$

where $\boldsymbol{\chi}$ is an inertially fixed unit vector (rotation axis) and ω is a scalar function of the time t. Hence the dynamical equations (Euler equations), appropriately generalized for the gyrostatic momentum, become [3]:

$$\sigma(\dot{\boldsymbol{\omega}}) + \boldsymbol{\omega} \wedge \sigma(\boldsymbol{\omega}) + \boldsymbol{\omega} \wedge \boldsymbol{I} = \eta \boldsymbol{c} \wedge \sigma(\boldsymbol{c}) \tag{1.1}$$

where σ denotes the central matrix of inertia of S, η is a scalar function (ever constant)[1]) depending on the time t by $r = |OG|$ and $\boldsymbol{c}$ is a unit vector: $\boldsymbol{c} = \text{vers } OG$ (orbital radius). Furthermore, the kinematical equations required for the complete determination of the motion are:

$$\dot{\boldsymbol{c}} = (\nu \boldsymbol{e} - \boldsymbol{\omega}) \wedge \boldsymbol{c}\,, \tag{1.2}$$

$$\dot{\boldsymbol{e}} = \boldsymbol{e} \wedge \boldsymbol{\omega} \tag{1.3}$$

where the dot denotes the derivative with respect to the body axes. The invariant relations of the system (1.2), (1.3) are expressed by:

$$\boldsymbol{e} \cdot \boldsymbol{c} = 0\,, \tag{1.4}$$
$$\boldsymbol{e} \cdot \boldsymbol{e} = 1\,, \tag{1.5}$$
$$\boldsymbol{c} \cdot \boldsymbol{c} = 1\,. \tag{1.6}$$

Without loss of generality, the motion equations can be written with respect to a body-fixed frame $R\Gamma(G;\, \boldsymbol{i}_1, \boldsymbol{i}_2, \boldsymbol{i}_3)$ where $\boldsymbol{i}_3 = \boldsymbol{\chi}$ and $\boldsymbol{i}_1$ and $\boldsymbol{i}_2$ are chosen in such a way that σ is expressed

[1]) We do not consider η = constant (circular orbit); this eventuality was studied in [4].

by the matrix:

$$\sigma = \begin{bmatrix} A & 0 & -B' \\ 0 & B & -A' \\ -B' & -A' & C \end{bmatrix}$$

The motions we are interested in are determined, in this frame, by the scalar functions ω, ν and η, and the components $e_i = e_i(t)$, $c_i = c_i(t)$, $(i = 1, 2, 3)$, of the vectorial functions $\mathbf{e}$, $\mathbf{c}$ in $R\Gamma$; obviously, e_3 does not depend upon t, namely $e_3 =$ constant.

2. All the possible stationary rotations

We can look for existence of such motions by the same method used in [4]. Since $\dot\omega = 0$, the projections in $R\Gamma$ of (1.1), (1.2), (1.3) and the invariant relations (1.4), (1.5), (1.6) look like:

$$A'\omega^2 - \omega I_2 = \eta[-A'c_3^2 + A'c_3^2 - B'c_1c_2 + (C - B)\, c_2c_3], \quad (2.1)$$

$$- B'\omega^2 + \omega I_1 = \eta[B'c_1^2 - B'c_3^2 + A'c_1c_2 + (A - C)\, c_1c_3], \quad (2.2)$$

$$(B - A)\, c_1c_2 - A'c_1c_3 + B'c_2c_3 = 0\,, \quad (2.3)$$

$$\dot c_1 = \nu e_2c_3 + c_2(\omega - \nu e_3)\,, \quad (2.4)$$

$$\dot c_2 = - \nu e_1c_3 - c_1(\omega - \nu e_3)\,, \quad (2.5)$$

$$\dot c_3 = \nu e_1c_2 - \nu e_2c_1\,, \quad (2.6)$$

$$\dot e_1 = \omega e_2\,, \quad (2.7)$$

$$\dot e_2 = - \omega e_1\,, \quad (2.8)$$

$$e_1c_1 + e_2c_2 + e_3c_3 = 0\,, \quad (2.9)$$

$$e_1^2 + e_2^2 = 1 - e_3^2\,, \quad (2.10)$$

$$c_1^2 + c_2^2 + c_3^2 = 1\,. \quad (2.11)$$

If we eliminate η by a linear combination between (2.1) and (2.2)[2], we obtain a new equation (2.12) independent of η. The equations (2.12), (2.3), (2.9), (2.10), (2.11), (the system Σ), define an algebraic variety in the five-dimensional real projective space determined by $R^5(e_1, e_2, c_1, c_2, c_3, z)$; this algebraic variety is the intersection of the corresponding hypersurfaces and a stationary rotation is associated with an algebraic variety of non-zero dimension, in fact with a proper real curve contained in the above intersection. Except for the case $A = B$, $A' = B' = 0$, which will be considered later, the intersection of Σ with the improper hyperplane in R^5 is easily seen to consist of the points:

$$P \equiv (1, i, 0, 0, 0, 0)\,, \quad \hat P \equiv (1, -i, 0, 0, 0, 0)$$

and

$$Q \equiv (0, 0, c_1, c_2, 1, 0)$$

where c_1 and c_2 are the possible solutions of the equations:

$$mB'c_1^2 + nA'c_2^2 + (mA' + nB')\, c_1c_2 + m(A - C)\, c_1 - n(C - B)\, c_2 - mB' - nA' = 0\,,$$

$$(B - A)\, c_1c_2 - A'c_1 + B'c_2 = 0\,,$$

$$c_1^2 + c_2^2 + 1 = 0\,,$$

with $m = A'\omega^2 - \omega I_2$ and $n = - B'\omega^2 + \omega I_1$. By the analogous considerations in [4], we obtain the following four cases:

1) $mB' + nA' = 0$,

2) $A' = 0$, $B'^2 + (B - A)(C - B) = 0$,

3) $B' = 0$, $A'^2 + (B - A)(A - C) = 0$[3],

4) $A' = B' = 0$.

Otherwise, it must be $e_1(t) = e_2(t) = c_3(t) = 0$ and so, by (2.3), either $A = B$ or $c_1(t) = 0$[4]. Both eventualities imply $\eta =$ constant directly by the equations. To discuss the cases 1), 2) and 4), we apply each condition to the equations of Σ. In each case, by (2.4), (2.5), (2.6), (2.7), (2.8) too, we obtain immediately $\omega = \nu e$; so the mass center G moves by a stationary motion and hence G describes a circular orbit. When $A = B$, $A' = B' = 0$, if it has not to be $\omega = \nu e$, the equation (2.12) gives $I_1 = I_2 = 0$ and so, by (2.1), (2.2), $c_3(t) = 0$. At last, all the equations are verified if and only if the following additional conditions hold:

$$e_1(t) = e_2(t) = 0,\ e_3^2 = 1\,,\ c_1^2 + c_2^2 = 1\,,\ \dot c_1 = c_2(\omega - \nu e_3)\,,$$
$$\dot c_2 = - c_1(\omega - \nu e_3)\,.$$

This case was mentioned in [3].

3. A class of non-stationary rotations

If we assume $A' = B' = 0$ and $c_3(t) = 0$, the three scalar equations, projections in $R\Gamma$ of (1.1), give:

$$I_1 = 0\,,\ I_2 = 0\,,\ C\dot\omega = \eta(B - A)\, c_1c_2\,.$$

By the equations (2.6), (2.9), (2.11), we obtain

$$e_1(t) = e_2(t) = 0\,,\ c_1^2 + c_2^2 = 1$$

and so by (2.10), (2.4), (2.5)

$$e_3^2 = 1,\ c_1 = c_2(\omega - \nu e_3),\ c_2 = - c_1(\omega - \nu e_3)\,.$$

If, furthermore, we assume either $A = C$ or $B = C$, we obtain the same above conditions; but if it is $A = B$, we obtain the previous stationary rotations.

4. Conclusions

Under a Newtonian force field, a „VOLTERRA-type" gyrostatic satellite in a KEPLER (non circular) orbit has not stationary rotations about its mass center. For a symmetric satellite they are only stationary rotations with the conditions mentioned in [3], and the gyrostatic momentum parallel to the rotation axis. A class of non-stationary rotations exist. It requires:

i) The gyrostatic momentum must be parallel to the rotation axis;

ii) The rotation axis must be a principal axis of inertia;

iii) The orbital plane must be a principal plane of inertia;

iv) The orbital radius has not to be parallel to a principal axis of inertia;

v) $C\dot\omega = \eta(B - A)\ c_1c_2$, $\dot c_1 = c_2(\omega - \nu e_3)$, $\dot c_2 = - c_1(\omega - \nu e_3)$, $e_3^2 = 1$.

For a symmetric satellite, it requires furthermore:

vi) The rotation axis has not to be parallel to the gyroscope axis.

References

1 LEVI CIVITA, T.; AMALDI, U., Lezioni di Meccanica Razionale, Zanichelli Bologna, 1927.
2 LEIMANIS, E., The General Problem of the Motion of Coupled Rigid Bodies About a Fixed Point, Springer Tracts in Natural Philosophy, Springer-Verlag Berlin/Heidelberg/New York, 1965.
3 GRIOLI, G., Particular Solutions in Stereodynamics; in: Stereodynamics, Corso C.I.M.E., Cremonese Roma, 1971.
4 BALLI, R.; NUCCI, M. C., Sulle rotazioni di un satellite girostatico in orbita circolare, Rend. Mat. Univ. Vol. 12-VI Roma, 1979.
5 BENTSIK, E., Sui possibili moti semplici di un satellite artificiale soggetto a forze newtoniane, Rend. Sem. Mat. Univ. Padova-39, 1967.

Eingereicht am 14. 11. 1979

Anschrift: Dr. MARIA CLARA NUCCI, C/O Istituto di Matematica, Via Vanvitelli 1, 06100 Perugia, Italia

[2]) If $\mathbf{c}$ is parallel to $\sigma(\mathbf{c})$, we have the same equations about a rigid satellite in a circular orbit [5].

[3]) The cases 2) and 3) are similar.

[4]) The case $c_2(t) = 0$ leads to the same conclusions about $c_1(t) = 0$.

ZAMM 60, 114 – 116 (1980)

D. HOMENTCOVSCHI

A New Form of the Fundamental Solution of Oseen Equations

The representation given by OSEEN in 1927 for the fundamental solution of OSEEN equations was, to our knowledge, the only one known hitherto. In the sequel, using the tool of the FOURIER transform we shall derive for the fundamental solution of OSEEN equations a representation which is expressed in terms of elementary functions only. This representation allows us to emphasize some properties of the fundamental solution.

1. In this note the OSEEN equations shall be written in the dimensionless form

$$\operatorname{div} \boldsymbol{V} = 0 \qquad (1)$$

$$\frac{\partial \boldsymbol{V}}{\partial x_1} + \operatorname{grad} p - \frac{1}{2\sigma} \Delta \boldsymbol{V} = 0$$

where 2σ stands for the REYNOLDS number. The fundamental solution of OSEEN equations consists of the tensor $\boldsymbol{t}$ (t_{jl}) and the vector $\boldsymbol{p}(p_j)$ $(j, l = 1, 2, 3)$ which satisfy the following system of equations

$$\sum_{l=1}^{3} \frac{\partial t_{jl}}{\partial x_l} = 0 \qquad (j = 1, 2, 3)$$

$$\frac{\partial t_{jl}}{\partial x_1} + \frac{\partial p_j}{\partial x_l} - \frac{1}{2\sigma} \Delta t_{jl} = \delta_{jl}\delta(\boldsymbol{x}) \qquad (j, l = 1, 2, 3) \qquad (2)$$

and which vanishes at infinity. The OSEEN representation [4] for the fundamental solution, taken over by other authors too [1], [2] is the following

$$t_{jl} = \frac{\sigma}{4\pi}\left(\Delta\varphi\delta_{jl} - \frac{\partial^2\varphi}{\partial x_j\,\partial x_l}\right) \qquad (j, l = 1, 2, 3)$$

$$p_j = -\frac{1}{8\pi}\frac{\partial}{\partial x_j}\left(\Delta\varphi - 2\sigma\frac{\partial\varphi}{\partial x_1}\right) \qquad (j = 1, 2, 3) \qquad (3)$$

$$\varphi(s) = \frac{1}{\sigma}\int_0^{\sigma s}\frac{1 - e^{-t}}{t}\,dt, \qquad s = r - x_1 \qquad (r = \sqrt{x_1^2 + x_2^2 + x_3^2})$$

In the form (3) the fundamental solution is difficult to use in applications. In the sequel we shall give a representation of the fundamental solution consisting of elementary functions only.

2. Taking the FOURIER transform of the system (2) with respect to the space variables we obtain the algebraic system

$$\sum_{l=1}^{3} ik_l \cdot T_{jl} = 0 \qquad (j = 1, 2, 3) \qquad (4)$$

$$ik_1 \cdot T_{jl} + ik_l \cdot P_j + \frac{1}{2\sigma}k^2 \cdot T_{jl} = \delta_{jl} \qquad (j, l = 1, 2, 3)\,.$$

The capital letters stand for FOURIER transforms ($k \equiv |\boldsymbol{k}| = \sqrt{k_1^2 + k_2^2 + k_3^2}$)

$$T_{jl}(\boldsymbol{k}) \equiv \mathcal{F}[t_{jl}(\boldsymbol{x})] = \iiint t_{jl}(\boldsymbol{x})\, e^{-i\boldsymbol{k}\cdot\boldsymbol{x}}\, d\boldsymbol{x}$$

$$P_j(\boldsymbol{k}) \equiv \mathcal{F}[p_j(\boldsymbol{x})] = \iiint p_j(\boldsymbol{x})\, e^{-i\boldsymbol{k}\cdot\boldsymbol{x}}\, d\boldsymbol{x}\,.$$

We can easily explicitate the solution of the system (4) in the form

$$P_j = -\frac{ik_j}{k^2} \qquad (j = 1, 2, 3) \qquad (5)$$

$$T_{jl} = \frac{2\sigma\delta_{jl}}{k^2 + 2\sigma i k_1} + \frac{ik_j \cdot ik_l}{ik_l}\left\{\frac{1}{k^2} - \frac{1}{k^2 + 2\sigma i k_1}\right\} \qquad (j, l = 1, 2, 3).$$

To calculate the inverse FOURIER transforms we shall use the relations

$$\mathcal{F}^{-1}\left[\frac{1}{k^2 + 2\sigma i k_1}\right] = \frac{e^{\sigma(x_1 - r)}}{4\pi r},$$

$$\int_{-\infty}^{x_1}\frac{\partial}{\partial x_\alpha}\frac{e^{\sigma(x_1 - r)}}{4\pi r}\,dx_1 = -\frac{1}{4\pi}\frac{x_\alpha}{x_2^2 + x_3^2}\left(1 + \frac{x_1}{r}\right)e^{\sigma(x_1 - r)} \qquad (6)$$

$$(\alpha = 2, 3)\,.$$

The first relation (6) may be obtained by using the results given in [3] and the second one can be checked directly.

Taking the FOURIER inverse transform from the relations (5) we obtain

$$p_j = -\frac{\partial}{\partial x_j}\frac{1}{4\pi r}$$

$$t_{j1} = 2\sigma\frac{e^{\sigma(x_1 - r)}}{4\pi r}\delta_{j1} + \frac{\partial}{\partial x_j}\left\{\frac{1 - e^{\sigma(x_1 - r)}}{4\pi r}\right\} \qquad (j = 1, 2, 3)$$

$$t_{\alpha\beta} = 2\sigma\frac{e^{\sigma(x_1 - r)}}{4\pi r}\delta_{\alpha\beta} - \frac{\partial}{\partial x_\alpha}\left\{\frac{1}{4\pi}\frac{x_\beta}{x_2^2 + x_3^2}\left(1 + \frac{x_1}{r}\right)(1 - e^{\sigma(x_1 - r)})\right\}$$

$$(\alpha, \beta = 2, 3)$$

$$t_{jl} = t_{lj} \qquad (j, l = 1, 2, 3)\,. \qquad (7)$$

Further on we shall give also the fundamental solution written in cylindrical coordinates $x_1 = x, \varrho, \theta$. We get

$$p_x = -\frac{\partial}{\partial x}\left(\frac{1}{4\pi r}\right), \quad p_\varrho = -\frac{\partial}{\partial \varrho}\left(\frac{1}{4\pi r}\right), \quad p_\theta = -\frac{1}{\varrho}\frac{\partial}{\partial\theta}\left(\frac{1}{4\pi r}\right) = 0,$$

$$t_{xx} = 2\sigma\frac{e^{\sigma(x - r)}}{4\pi r} + \frac{\partial}{\partial x}\left\{\frac{1 - e^{\sigma(x - r)}}{4\pi r}\right\},$$

$$t_{x\varrho} = t_{\varrho x} = \frac{\partial}{\partial\varrho}\left\{\frac{1 - e^{\sigma(x - r)}}{4\pi r}\right\}, \qquad t_{\theta x} = t_{x\theta} = 0 \qquad (8)$$

$$t_{\varrho\varrho} = 2\sigma\frac{e^{\sigma(x - r)}}{4\pi r} - \frac{\partial}{\partial\varrho}\left\{\frac{\sigma\varrho}{4\pi r}\frac{\operatorname{sh}\left(\frac{\sigma}{2}(x - r)\right)}{\frac{\sigma}{2}(x - r)}e^{\frac{\sigma}{2}(x - r)}\right\},$$

$$t_{\theta\theta} = 2\sigma\frac{e^{\sigma(x - r)}}{4\pi r} - \frac{\sigma}{4\pi r}\frac{\operatorname{sh}\frac{\sigma}{2}(x - r)}{\frac{\sigma}{2}(x - r)}e^{\frac{\sigma}{2}(x - r)}, \qquad t_{\theta\varrho} = t_{\varrho\theta} = 0\,.$$

3. In the following we shall emphasize some properties of the fundamental solution of OSEEN equations which follow from the representation (7).

a) In the limit case of an ideal fluid ($\sigma \to \infty$) we get

$$t_{j1} = \frac{\partial}{\partial x_j}\left(\frac{1}{4\pi r}\right) \qquad (j = 1, 2, 3)$$

$$t_{\alpha\beta} = -\frac{\partial}{\partial x_\alpha}\left\{\frac{1}{4\pi}\frac{x_\beta}{x_2^2 + x_3^2}\left(1 + \frac{x}{r}\right)\right\} \qquad (\alpha, \beta = 2, 3) \qquad (7')$$

and, in cylindrical coordinates,

$$t_{xx} = \frac{\partial}{\partial x}\left(\frac{1}{4\pi r}\right), \quad t_{x\varrho} = t_{\varrho x} = \frac{\partial}{\partial\varrho}\left(\frac{1}{4\pi r}\right), \quad t_{\theta x} = t_{x\theta} = 0,$$

$$t_{\varrho\varrho} = -\frac{\partial}{\partial\varrho}\left\{\frac{1}{4\pi\varrho}\left(1 + \frac{x_1}{r}\right)\right\}; \qquad t_{\theta\theta} = -\frac{1}{4\pi\varrho^2}\left(1 + \frac{x}{r}\right) \qquad (8')$$

for all directions but $\varrho = 0$, $x > 0$.

The representation (7′) of the solution coincides with that used to build the theory of the lifting surface from the case of ideal fluids.

b) The fundamental solution of OSEEN equation is vanishing as $r \to \infty$ along all the direction. As $x \to -\infty$ the solution vanishes due to the factor $(1 + x/r)$ and at infinity in aval — due to the factor $(1 - \exp\{\sigma(x - r)\})$. In the case of the ideal fluid, the relations (7′) show the fact that the fundamental solution vanishes, at big distances, everywhere but the TREFFTZ's plane ($x \to +\infty$), where nevertheless it is bounded.

c) From the relations (8) it follows that the fundamental solution is a regular function in all the plane but the origine. At the origine it behaves as $1/r$.

In the case of the ideal fluid from the relations (8′) it follows that the fundamental solution has a singularity on the half-axis $\varrho = 0$, $x > 0$; this singularity is modeling the "free vortices" from the theory of lifting wing.

d) For small values of σ we have from (7)

$$t_{ij} = \frac{\sigma}{4\pi}\left\{\frac{\delta_{ij}}{r} + \frac{x_i x_j}{r^3}\right\} + 0(\sigma^2)\,.$$

Thus, to σ^2 order, the fundamental solution of the OSEEN equations coincides with the fundamental solution of the STOKES equation.

References

1 BERKER, R., Intégration des équations du mouvement d'un fluide visqueux incompressible, Handbuch der Physik VIII/2, Springer Verlag (1963).
2 CHANG, I. D., FINN, R., On the Solution of a Class of Equations Occuring in Continuum Mechanics, with Applications to the STOKES Paradox, Arch. Rat. Mech. Anal. 8 (1962) pp. 388–401.
3 GELFAND, I. M.; SHILOV, G. E., Generalized Functions, Academic Press (1964).
4 OSEEN, C. W., Neuere Methoden und Ergebnisse in der Hydrodynamik, Leipzig, Akademische Verlagsgesellschaft 1927.

Eingereicht am 5. 2. 1978

Anschrift: Dr. DOREL HOMENTCOVSCHI, Central Institute of Mathematics, Academiei Street, 14, Bucharest 1, Romania

ZAMM 60, 116–117 (1980)

V. D. SHARMA

A Note on the Propagation of Sonic Waves in Radiating Gases

In this work the growth and decay properties of sonic waves discussed in [1] are modified. In addition to the cases of plane and spherical waves treated in [1], this paper also discusses waves with cylindrical geometry. We find that a compressive wave terminates into a shock after a finite time only if the magnitude of the initial discontinuity associated with the wave exceeds a critical value. It is shown that a compressive wave decays completely or takes a stable wave form according as the initial discontinuity associated with the wave is less than or equal to this critical value. The existence of such critical values is an interesting physical result which has not been mentioned in [1]. The effects of wave front curvature on the growth and decay properties of these waves which were not considered in [1] are also discussed here. It is found that the wave front curvature as well as the radiative transfer both have a delaying effect on the formation of a shock wave.

1. Introduction

In a recent paper SRINIVASAN and RAM [1], following the singular surface theory, have studied the propagation of sonic waves in a radiating gas. The purpose of this paper is to discuss the equation

$$\zeta = \frac{\zeta_0 (1 - 2\Omega_0\sigma + K_0\sigma^2)^{-1/2} \exp(-A\sigma)}{1 + \left(\frac{\nu+1}{2\varrho}\right)\zeta_0 \int\limits_0^\sigma (1 - 2\Omega_0\sigma + K_0\sigma^2)^{-1/2} \exp(-A\sigma)\, d\sigma} \tag{1}$$

(equation (17) of their paper) for waves with plane, cylindrical and spherical geometry. The symbols appearing in (1) are exactly the same as in their paper; $\sigma = G(t - t_0)$ is the distance measured from the initial position of the wave surface $\Sigma(t)$ along its normal trajectories, $G = (\nu p/\varrho)^{1/2}$ is the sonic speed of propagation of the discontinuity surface $\Sigma(t)$, ζ_0 is the value of the discontinuity ζ at the initial time $t = t_0$, Ω_0 and K_0 are respectively the mean and GAUSSian curvatures of $\Sigma(t_0)$, ν is the specific heat ratio and A is a positive constant given by $A = \frac{2(\nu-1)^2 Q}{G^3}$, where $Q = 4\alpha_p a_R T^4$ is the rate of energy loss by the gas per unit volume.

2. Behaviour at the Wave Front

Case I. Plane Waves

For a plane wave front $\Omega_0 = 0 = K_0$, the equation (1) reduces to the form

$$\zeta = \frac{\zeta_0 \exp(-A\sigma)}{1 + \frac{\zeta_0}{\Lambda}\{1 - \exp(-A\sigma)\}} \tag{2}$$

where $\Lambda = 2A\varrho/(\nu + 1)$.

Equation (2) shows that if $\zeta_0 < 0$ (i.e. a compression wave front) and $|\zeta_0| < \Lambda$, then $\zeta \to 0$ as $\sigma \to \infty$, i.e. a sonic wave of compressive nature decays and damps out ultimately. Further if $\zeta_0 < 0$ and $|\zeta_0| = \Lambda$, then $\zeta = \zeta_0$, and the wave takes a stable wave form. But, if $\zeta_0 < 0$ and $|\zeta_0| > \Lambda$, then $|\zeta| \to \infty$ for a finite σ_c^* given by

$$\sigma_c^* = \frac{1}{A} \log\left(1 - \frac{\Lambda}{|\zeta_0|}\right)^{-1}. \tag{3}$$

Thus at a finite time t^* given by $t^* = t_0 + \sigma_c^*/G$, a plane compressive wave steepens up into a shock. Here, we find that the initial value of the discontinuity given by $\zeta_0 = -\Lambda$ plays an important role in determining whether a compressive wave shall grow or decay. Accordingly we designate the value $\zeta_0 = -\Lambda$ as the critical ζ_0 for the plane wave. It is evident from the expressions of Λ and t^* that they are increasing functions of A, and hence we conclude that the radiating flux causes the shock formation time to increase.

Case II. Spherical Waves

If the wave front $\Sigma(t)$ at time t_0 is a sphere of radius R_0, then at any time $t > t_0$, $\Sigma(t)$ is a sphere of radius $R = R_0 + G(t - t_0)$. Also, for this case we have $\Omega_0 = -1/R_0$, $K_0 = 1/R_0^2$ and $\sigma = R - R_0$. Then (1) becomes

$$\zeta = \frac{\zeta_0(R_0/R)\exp(-A(R - R_0))}{1 + \left(\frac{\nu+1}{2\varrho}\right)\zeta_0 R_0 E_i(AR_0)\left\{1 - \frac{E_i(AR)}{E_i(AR_0)}\right\}\exp(AR_0)} \tag{4}$$

where $E_i(x) = \int\limits_x^\infty t^{-1} e^{-t}\, dt$ is a tabulated function known as the exponential integral function. The term in the curly bracket in the denominator of (4) increases monotonically from 0 to 1 as R increases from R_0 to ∞. Hence, in this case also there exists a critical value of initial discontinuity ζ_c given by

$$\zeta_c = -\frac{2\varrho \exp(-AR_0)}{(\nu + 1) R_0 E_i(AR_0)}$$

such that if $\zeta_0 < 0$ and $|\zeta_0| < |\zeta_c|$, then $\zeta \to 0$ as $R \to \infty$, the wave decays and damps out ultimately. Further, if $\zeta_0 = \zeta_c$, then (4) shows that $\zeta \to -\Lambda$ (critical ζ_0 for plane wave) as $R \to \infty$, i.e. the wave does not completely decay and ultimately takes a stable wave form. But, if $\zeta_0 < 0$ and $|\zeta_0| > |\zeta_c|$, then we have the criteria

$$E_i(AR^*/G) = \left(1 - \frac{|\zeta_c|}{|\zeta_0|}\right) E_i(AR_0/G) \tag{5}$$

for the shock formation at a finite $R = R^*$, where $R^* = R_0 + G(t^* - t_0)$. From the expression of ζ_c, using the inequality $E_i(x) < e^{-x}/x$, it follows that the magnitude of the critical value of the initial discontinuity for a spherical wave is greater than the corresponding value for a plane wave i.e. $|\zeta_c| > \Lambda$. It is also evident from the expression of ζ_c that $\frac{\partial|\zeta_c|}{\partial A} > 0$, which means that the magnitude of the critical value of the initial discontinuity increases with A. Also $\frac{\partial|\zeta_c|}{\partial R_0} < 0$, which implies that the initial curvature has a stabilizing effect on the tendency of the wave surface $\Sigma(t)$ to grow into a shock in the sense that an increase in the value of the initial curvature causes an increase in the critical value $|\zeta_c|$. Further, it is also evident from (5) that $\frac{\partial t^*}{\partial A} > 0$, which means that an increase in A will cause the shock formation time t^* to increase. On the other hand if the wave is expensive ($\zeta_0 > 0$), then it follows from (4) that $\zeta \to 0$ as $R \to \infty$, the wave decays and damps out ultimately.

Case III. Cylindrical Waves

If the expanding wave front $\Sigma(t)$ at $t = t_0$ is a cylinder of radius R_0, then at any time $t > t_0$, $\Sigma(t)$ is a cylinder of radius $R = R_0 + G(t - t_0)$. For this case we have, $\Omega_0 = -1/2R_0$, $K_0 = 0$, $\sigma = R - R_0$ and (1) reduces to

$$\zeta = \frac{\zeta_0(R_0/R)^{1/2}\exp(-A(R - R_0))}{1 + \frac{(\nu+1)}{2\varrho}\zeta_0(\pi R_0/A)^{1/2}\exp(AR_0)\,\mathrm{erfc}(AR_0)^{1/2}\left\{1 - \frac{\mathrm{erfc}(AR)^{1/2}}{\mathrm{erfc}(AR_0)^{1/2}}\right\}} \tag{6}$$

where $\operatorname{erfc}(x) = \frac{2}{\sqrt{\pi}} \int_x^{\infty} e^{-t^2}\, dt$ is the complementary error function. Now if $\zeta_0 > 0$, or if $\zeta_0 < 0$ and $|\zeta_0| < |\hat{\zeta}_c|$ where

$$\hat{\zeta}_c = \frac{-2\varrho(\pi R_0/A)^{-1/2} \exp(-A R_0)}{(\nu + 1) \operatorname{erfc}(A R_0)^{1/2}},$$

then, it follows from (6) that $\zeta \to 0$ as $R \to \infty$, the wave damps out. Further if $\zeta_0 = \hat{\zeta}_c$, then (6) shows that the wave decays and $\zeta \to -A$ (critical value of the initial discontinuity for a plane wave) as $R \to \infty$. i.e. the wave ultimately takes a stable wave form. From expression of $\hat{\zeta}_c$, using the inequality $\operatorname{erfc}(x) < e^{-x^2}/x\sqrt{\pi}$, it follows immediately that the magnitude of critical value $\hat{\zeta}_c$ for a cylindrical wave is greater than the corresponding value for a plane wave. But if $\zeta_0 < 0$ and $|\zeta_0| > |\hat{\zeta}_c|$, then we have the criteria

$$\operatorname{erfc}(A\hat{R})^{1/2} = \left(1 - \frac{|\hat{\zeta}_c|}{|\zeta_0|}\right) \operatorname{erfc}(A R_0)^{1/2} \tag{7}$$

for the shock formation at a finite $R = \hat{R}$, where $\hat{R} = R_0 + G(\hat{t} - t_0)$. It is evident from the expression of $\hat{\zeta}_c$ and (7) that the wave front curvature as well as the radiating flux both cause the shock formation time $\hat{t}$ to increase.

References

1 SRINIVASAN, S., and RAM, R., Propagation of Sonic Waves in Radiating Gases, ZAMM 57, 191–193 (1977).

Eingereicht am 9. 12. 1978

Anschrift: Dr. V. D. SHARMA, Applied Mathematics Section, Institute of Technology, Banaras Hindu University, Varanasi, India

ZAMM 60, 117–118 (1980)

O. BECKER

Eine Theorie dritter Ordnung des räumlich gekrümmten Trägers

1. Problemstellung

Bei der korrekten Formulierung von Stabilitätsproblemen ist i. a. von einer Theorie dritter Ordnung auszugehen. Für den schwach gekrümmten Träger mit offenem Querschnitt wird in [1] unter Berücksichtigung von Querkraft- und Wölbdeformationen eine solche Theorie entwickelt.

Hier wird über eine Theorie dritter Ordnung für den räumlich beliebig gekrümmten Träger mit nichtoffenem Querschnitt berichtet; die Ableitung der Beziehungen findet sich in [2]. Verschiebungen und Drehungen der Trägerquerschnitte sind endliche Größen, wobei Querkraftdeformationen in linearer Näherung berücksichtigt und Verwölbungen vernachlässigt werden. Der Werkstoff ist linear-elastisch; die Spannungen senkrecht zur Trägerachse sind vernachlässigbar.

2. Verzerrungszustand

Als Basisvektoren werden die $\boldsymbol{e}_{i0}$ gewählt, die sich auf die Tangente ($\boldsymbol{e}_{10}$) an die undeformierte Trägerachse $\boldsymbol{r}_0(s_0)$ mit Krümmung $\varkappa_0$ und Windung τ_0 und auf die Hauptträgheitsachsen ($\boldsymbol{e}_{20}, \boldsymbol{e}_{30}$) des Querschnitts beziehen. Der Krümmungsvektor der undeformierten Trägerachse ist durch (1) gegeben; ψ_0 ist der Winkel zwischen $\boldsymbol{e}_{20}$ und der Hauptnormalenrichtung. Ortsvektor und Gittervektoren des undeformierten Trägers sind durch (2) bis (4) bestimmt; (5) stellt die Koeffizienten für die Transformation der kovarianten Koordinaten eines Tensors aus den $\boldsymbol{g}^{k0}$ in die orthonormale Basis $\boldsymbol{e}^{k0} = \boldsymbol{e}_{k0}$ dar.

$$\boldsymbol{k} = k_{i0}\boldsymbol{e}_{i0}; \quad k_{10} = \tau_0 + \frac{\partial \psi_0}{\partial s_0}, \quad k_{20} = \varkappa_0 \sin\psi_0, \quad k_{30} = \varkappa_0 \cos\psi_0, \tag{1}$$

$$\boldsymbol{R}_0 = \boldsymbol{r}_0 + x_2\boldsymbol{e}_{20} + x_3\boldsymbol{e}_{30}, \tag{2}$$

$$\boldsymbol{g}_{10} = \sqrt{g_0}\,\boldsymbol{e}_{10} - k_{10}x_3\boldsymbol{e}_{20} + k_{10}x_2\boldsymbol{e}_{30}, \quad \boldsymbol{g}_{20} = \boldsymbol{e}_{20}, \quad \boldsymbol{g}_{30} = \boldsymbol{e}_{30}, \tag{3}$$

$$\sqrt{g_0} = 1 + k_{20}x_3 - k_{30}x_2, \tag{4}$$

$$\{\alpha_k^{i}\} = \begin{bmatrix} 1/\sqrt{g_0} & k_{10}x_3/\sqrt{g_0} & -k_{10}x_2/\sqrt{g_0} \\ 0 & 1 & 0 \\ 0 & 0 & 1 \end{bmatrix}. \tag{5}$$

Bei Deformation des Trägers geht $\boldsymbol{r}_0$ in die deformierte Trägerachse (6) über und die fest mit ihr verbunden zu denkenden $\boldsymbol{e}_{i0}$ drehen sich in eine neue Lage $\boldsymbol{e}_i$ gemäß (7). Die Koordinaten des differentiellen Drehvektors für diese endliche Drehung sind in den $\boldsymbol{e}_i$ durch (8) darstellbar. Die Koeffizienten t_{ik}, p_{ij} hängen von der Wahl der drei nacheinander auszuführenden, endlichen Teildrehungen φ_j ab; vergl. [2].

$$\boldsymbol{r} = \boldsymbol{r}_0 + \boldsymbol{u} \quad \text{mit} \quad \boldsymbol{u} = u_{i0}\boldsymbol{e}_{i0}, \tag{6}$$

$$\boldsymbol{e}_i = t_{ik}\boldsymbol{e}_{k0}, \tag{7}$$

$$\mathrm{d}\Phi_i = p_{ij}\,\mathrm{d}\varphi_j \quad \text{mit} \quad \mathrm{d}\boldsymbol{\varphi} = \mathrm{d}\Phi_i\,\boldsymbol{e}_i. \tag{8}$$

Gleichzeitig geht $\mathrm{d}s_0$ in das Bogenelement $\mathrm{d}s = (1 + \varepsilon_1)\,\mathrm{d}s_0$ und $\boldsymbol{k}$ in den Krümmungsvektor $\boldsymbol{\varkappa} = \varkappa_i\boldsymbol{e}_i$ der deformierten Trägerachse über. Die Koordinaten $\varkappa_i$ lassen sich gegenüber der Darstellung $\varkappa_i = (\partial\boldsymbol{e}_j/\partial s)\cdot\boldsymbol{e}_k$, $i, j, k = 1, 2, 3$ zykl., wesentlich einfacher durch (9) berechnen; analog die Koordinaten des Vektors der Winkelgeschwindigkeit $\boldsymbol{\omega} = \omega_i\boldsymbol{e}_i$. Die Dehnung ε_1 und die effektiven Krümmungsänderungen ψ_i der Trägerachse sind durch (10), (11) bestimmt. Den sechs Deformationsgrößen u_{i0}, φ_j stehen vier Freiheitsgrade der $\boldsymbol{e}_2\boldsymbol{e}_3$-Ebene gegenüber, so daß die Zwangsbedingungen (12) bestehen. Mit $\boldsymbol{\varkappa}$, $\boldsymbol{\omega}$ gelten die Ableitungsgleichungen (13), wobei ein Stern die Differentiation bei konstanten $\boldsymbol{e}_i$ kennzeichnet.

$$\varkappa_i = \left(t_{ij}k_{j0} + p_{ij}\frac{\partial\varphi_j}{\partial s_0}\right)\Big/(1 + \varepsilon_1), \quad \omega_i = p_{ij}\frac{\partial\varphi_j}{\partial t}; \tag{9}$$

$$\varepsilon_1 = \sqrt{1 + 2e + e^2 + d_2^2 + d_3^2} - 1, \quad \psi_i = (1 + \varepsilon_1)\varkappa_i - k_{i0}; \tag{10}$$

$$e = \frac{\partial u_{10}}{\partial s_0} + k_{20}u_{30} - k_{30}u_{20},$$

$$d_3 = \frac{\partial u_{20}}{\partial s_0} + k_{30}u_{10} - k_{10}u_{30}, \tag{11}$$

$$-d_2 = \frac{\partial u_{30}}{\partial s_0} + k_{10}u_{20} - k_{20}u_{10};$$

$$t_{11} = (1 + e)/(1 + \varepsilon_1), \quad t_{12} = d_3/(1 + \varepsilon_1), \quad t_{13} = -d_2/(1 + \varepsilon_1)$$
$$\text{mit } t_{1i}t_{1i} = 1; \tag{12}$$

$$\frac{\partial\boldsymbol{a}}{\partial s} = \frac{\partial^*\boldsymbol{a}}{\partial s} + \boldsymbol{\varkappa}\times\boldsymbol{a}, \quad \frac{\partial\boldsymbol{a}}{\partial t} = \frac{\partial^*\boldsymbol{a}}{\partial t} + \boldsymbol{\omega}\times\boldsymbol{a}. \tag{13}$$

Beim schubstarren Träger ist $\mathring{\boldsymbol{e}}_i = \boldsymbol{e}_i$, wenn die $\mathring{\boldsymbol{e}}_i$ die neue Lage kennzeichnen, in welche die fest mit dem Querschnitt verbunden zu denkenden $\boldsymbol{e}_{i0}$ übergehen. Beim schubsteifen Träger steht der Querschnitt nicht mehr senkrecht zur deformierten Trägerachse, $\mathring{\boldsymbol{e}}_i \neq \boldsymbol{e}_i$. Für die Querkraftdeformationen $\hat{u}_{i0}$, $\hat{\varphi}_j$ gilt in linearer Näherung (14), (15). Zu (7) bis (12) analoge Beziehungen gelten auch für die mit einem *o* gekennzeichneten Größen.

$$u_{i0} = \mathring{u}_{i0} + \hat{u}_{i0} \quad \text{mit} \quad \hat{u}_{10} = 0, \quad \varphi_j = \mathring{\varphi}_j + \hat{\varphi}_j \quad \text{mit} \quad \hat{\varphi}_1 = 0; \tag{14}$$

$$\mathring{\boldsymbol{e}}_1 = \boldsymbol{e}_1 - \hat{\varphi}_3\boldsymbol{e}_2 + \hat{\varphi}_2\boldsymbol{e}_3, \quad \mathring{\boldsymbol{e}}_2 = \boldsymbol{e}_2 + \hat{\varphi}_3\boldsymbol{e}_1, \quad \mathring{\boldsymbol{e}}_3 = \boldsymbol{e}_3 - \hat{\varphi}_2\boldsymbol{e}_1. \tag{15}$$

Ortsvektor und Gittervektoren des deformierten Trägers sind durch (16) bis (18) gegeben. Die Koordinaten des GREENschen Verzerrungstensors in der kontravarianten Basis $\boldsymbol{g}^{k0}$ transformieren sich mit (5) in die Koordinaten e_{ij} in der orthonormalen Basis $\boldsymbol{e}_{k0}$; sie sind durch (19), (20) bestimmt.

$$\boldsymbol{R} = \boldsymbol{r} + x_2\mathring{\boldsymbol{e}}_2 + x_3\mathring{\boldsymbol{e}}_3, \tag{16}$$

$$\boldsymbol{g}_1 = \sqrt{g}\,\mathring{\boldsymbol{e}}_1 - [(k_{10} + \mathring{\psi}_1)x_3 - \hat{\varphi}_3]\,\mathring{\boldsymbol{e}}_2 + [(k_{10} + \mathring{\psi}_1)x_2 - \hat{\varphi}_2]\,\mathring{\boldsymbol{e}}_3,$$
$$\boldsymbol{g}_2 = \mathring{\boldsymbol{e}}_2, \quad \boldsymbol{g}_3 = \mathring{\boldsymbol{e}}_3; \tag{17}$$

$$\sqrt{g} = \sqrt{g_0} + \varepsilon_1 + \mathring{\psi}_2x_3 - \mathring{\psi}_3x_2, \quad \mathring{\psi}_i = (1 + \mathring{\varepsilon}_1)\mathring{\varkappa}_i - k_{i0}; \tag{18}$$

$$e_{11} = \varepsilon_{11} + \tfrac{1}{2}\,\varepsilon_{11}^2 + \tfrac{1}{2}\,\psi_1^2(x_2^2 + x_3^2)/g_0\,,$$
$$e_{12} = e_{21} = \tfrac{1}{2}\,\varepsilon_{12}\,, \quad e_{13} = e_{31} = \tfrac{1}{2}\,\varepsilon_{13}\,, \tag{19}$$
$$e_{22} = e_{23} = e_{32} = e_{33} = 0;$$
$$\varepsilon_{11} = (\varepsilon_1 + \mathring{\psi}_2 x_3 - \mathring{\psi}_3 x_2)/\sqrt{g_0}\,,$$
$$\varepsilon_{12} = (\hat{\varphi}_3 - \mathring{\psi}_1 x_3)/\sqrt{g_0}\,, \tag{20}$$
$$e_{13} = (-\hat{\varphi}_2 + \mathring{\psi}_1 x_2)/\sqrt{g_0}\,.$$

3. Schnittgrößen

Im Rahmen einer eindimensionalen Theorie führen die Spannungen senkrecht zur Trägerachse zu Widersprüchen an der lastfreien Oberfläche des Trägers. Sie werden daher vernachlässigt, so daß sich die Spannungsgrößen (21) und die Variation (22) der im Trägerelement gespeicherten Formänderungsenergie ergeben. Die in bezug auf den Querschnitt im deformierten Zustand formulierten Schnittkräfte $\boldsymbol{F} = F_i \mathring{\boldsymbol{e}}_i$ und Schnittmomente $\boldsymbol{M} = M_i \mathring{\boldsymbol{e}}_i$ leisten die virtuelle Arbeit (23). Aus der Gleichheit von (22), (23) folgen die Schnittgrößen (24), (25).

$$k^{11} = Ee_{11}\,, \quad k^{12} = k^{21} = 2Ge_{12}\,, \quad k^{13} = k^{31} = 2Ge_{13}\,,$$
$$k^{22} = k^{23} = k^{32} = k^{33} = 0; \tag{21}$$
$$\mathrm{d}(\delta W) = \int_A [Ee_{11}\delta e_{11} + 4G(e_{12}\delta e_{12} + e_{13}\delta e_{13})]\sqrt{g_0}\,\mathrm{d}A\,\mathrm{d}s_0; \tag{22}$$
$$\mathrm{d}(\delta A_S) = (F_1\delta\varepsilon_1 + F_2\delta\hat{\varphi}_3 - F_3\delta\hat{\varphi}_2 + M_1\delta\mathring{\psi}_1 + M_2\delta\mathring{\psi}_2 + M_3\delta\mathring{\psi}_3)\,\mathrm{d}s_0; \tag{23}$$
$$F_1 = E\int_A e_{11}(1 + \varepsilon_{11})\,\mathrm{d}A\,, \quad F_2 = G\int_A \varepsilon_{12}\,\mathrm{d}A\,,$$
$$F_3 = G\int_A \varepsilon_{13}\,\mathrm{d}A; \tag{24}$$
$$M_1 = G\int_A (\varepsilon_{13}x_2 - \varepsilon_{12}x_3)\,\mathrm{d}A + E\int_A e_{11}\mathring{\psi}_1(x_2^2 + x_3^2)\frac{1}{\sqrt{g_0}}\,\mathrm{d}A\,,$$
$$M_2 = E\int_A e_{11}(1 + \varepsilon_{11})\,x_3\,\mathrm{d}A\,, \quad M_3 = -E\int_A e_{11}(1 + \varepsilon_{11})\,x_2\,\mathrm{d}A\,. \tag{25}$$

4. Bewegungsgleichungen und Randaussagen

Die Koordinaten eines Vektors in den $\boldsymbol{e}_i$ bzw. $\mathring{\boldsymbol{e}}_i$ unterscheiden sich nur durch vernachlässigbar kleine Größen, so daß im weiteren $\boldsymbol{e}_i \approx \mathring{\boldsymbol{e}}_i$ gesetzt werden kann. Bewegungsgleichungen und Randaussagen folgen aus dem Prinzip von HAMILTON (26) mit dem Impuls- und Drehimpulsvektor gemäß (27) bis (30), den Streckenlasten $\boldsymbol{f}$ und $\boldsymbol{m}$ sowie den Randlasten $\boldsymbol{F}_A$, $\boldsymbol{M}_A$, $\boldsymbol{F}_E$, $\boldsymbol{M}_E$, in denen nach Anwendung des Befreiungsprinzips von LAGRANGE auch die Reaktionen infolge vorgegebener Bindungen enthalten sind. Die in (26) auftretenden Variationen sind durch (31), (32) gegeben.

$$\int_{t_1}^{t_2}\Big[\int_{S_0}(\boldsymbol{I}\cdot\delta\boldsymbol{v} + \boldsymbol{B}\cdot\delta\mathring{\boldsymbol{\omega}} - \boldsymbol{F}\cdot\delta\boldsymbol{\alpha} - \boldsymbol{M}\cdot\delta\check{\boldsymbol{\psi}} + \boldsymbol{f}\cdot\delta\boldsymbol{u} + \boldsymbol{m}\cdot\delta\mathring{\boldsymbol{\varphi}})\,\mathrm{d}s_0 + \boldsymbol{F}_A\cdot\delta\boldsymbol{u}_A + \boldsymbol{M}_A\cdot\delta\mathring{\boldsymbol{\varphi}}_A + \boldsymbol{F}_E\cdot\delta\boldsymbol{u}_E + \boldsymbol{M}_E\cdot\delta\mathring{\boldsymbol{\varphi}}_E\Big]\,\mathrm{d}t = 0; \tag{26}$$
$$\boldsymbol{I} = \varrho(\mathfrak{A}\boldsymbol{v} + \mathring{\boldsymbol{\omega}}\times\mathfrak{S})\,, \quad \boldsymbol{B} = \varrho(\mathfrak{S}\times\boldsymbol{v} + \mathfrak{J}\cdot\mathring{\boldsymbol{\omega}}); \tag{27}$$
$$\mathfrak{A} = \int_A \sqrt{g_0}\,\mathrm{d}A\,, \quad \mathfrak{S} = \mathfrak{S}_i\mathring{\boldsymbol{e}}_i\,, \quad \mathfrak{J} = \mathfrak{J}_{ik}\mathring{\boldsymbol{e}}_i\mathring{\boldsymbol{e}}_k; \tag{28}$$
$$\mathfrak{S}_1 = 0\,, \qquad \mathfrak{S}_2 = \int_A x_2\sqrt{g_0}\,\mathrm{d}A\,, \quad \mathfrak{S}_3 = \int_A x_3\sqrt{g_0}\,\mathrm{d}A; \tag{29}$$
$$\mathfrak{J}_{22} = \int_A x_3^2\sqrt{g_0}\,\mathrm{d}A\,, \quad \mathfrak{J}_{23} = \mathfrak{J}_{32} = -\int_A x_2x_3\sqrt{g_0}\,\mathrm{d}A\,,$$
$$\mathfrak{J}_{33} = \int_A x_2^2\sqrt{g_0}\,\mathrm{d}A\,, \quad \mathfrak{J}_{11} = \mathfrak{J}_{22} + \mathfrak{J}_{33}\,, \tag{30}$$
$$\mathfrak{J}_{12} = \mathfrak{J}_{21} = \mathfrak{J}_{13} = \mathfrak{J}_{31} = 0;$$
$$\delta\boldsymbol{u} = \delta U_i\boldsymbol{e}_i\,, \quad \delta\boldsymbol{v} = \frac{\partial}{\partial t}(\delta\boldsymbol{u})\,,$$
$$\delta\boldsymbol{\alpha} = (1 + \varepsilon_1)\left[\frac{\partial}{\partial s}(\delta\boldsymbol{u}) + \boldsymbol{e}_1\times\delta\mathring{\boldsymbol{\varphi}}\right]; \tag{31}$$
$$\delta\mathring{\boldsymbol{\varphi}} = \delta\mathring{\Phi}_i\boldsymbol{e}_i\,, \quad \delta\mathring{\boldsymbol{\omega}} = \frac{\partial}{\partial t}(\delta\mathring{\boldsymbol{\varphi}})\,, \quad \delta\check{\boldsymbol{\psi}} = (1 + \varepsilon_1)\frac{\partial}{\partial s}(\delta\mathring{\boldsymbol{\varphi}})\,. \tag{32}$$

Aus der Willkürlichkeit und Unabhängigkeit der Koordinaten δU_i, $\delta\mathring{\Phi}_j$ folgen aus (26) bis (32) die Bewegungsgleichungen (33), (34), die das Kräfte- und Momentengleichgewicht am verformten Trägerelement ausdrücken, und die Randaussagen (35), (36), welche Gleichgewichtsbedingungen zwischen Schnittgrößen, äußeren Lasten und Reaktionen an den Trägerenden s_A, s_E für den verformten Zustand darstellen. Aus (35), (36) folgen die Bestimmungsgleichungen für die Reaktionen bei vorgegebenen Bindungen bzw. die natürlichen Randbedingungen.

$$(1 + \varepsilon_1)\frac{\partial\boldsymbol{F}}{\partial s} + \boldsymbol{f} - \frac{\partial\boldsymbol{I}}{\partial t} = \boldsymbol{0}; \tag{33}$$
$$(1 + \varepsilon_1)\frac{\partial\boldsymbol{M}}{\partial s} + (1 + \varepsilon_1)\,\boldsymbol{e}_1\times\boldsymbol{F} + \boldsymbol{m} - \frac{\partial\boldsymbol{B}}{\partial t} = \boldsymbol{0}; \tag{34}$$
$$\boldsymbol{F}(s_A) + \boldsymbol{F}_A = \boldsymbol{0}\,, \quad \boldsymbol{M}(s_A) + \boldsymbol{M}_A = \boldsymbol{0}; \tag{35}$$
$$-\boldsymbol{F}(s_E) + \boldsymbol{F}_E = \boldsymbol{0}\,, \quad -\boldsymbol{M}(s_E) + \boldsymbol{M}_E = \boldsymbol{0}; \tag{36}$$

5. Zusammenfassung

Schnittgrößen, Impuls- und Drehimpulsvektor sind in bezug auf den Querschnitt im deformierten Zustand des Trägers definiert; die Gleichgewichtsaussagen am Trägerelement und an den Trägerenden beziehen sich ebenfalls auf den deformierten Zustand; in den Verzerrungen der Trägerachse, in den Schnittgrößen und Trägheitsgliedern werden sämtliche nichtlinearen Terme beibehalten. Damit stellen die erhaltenen Ergebnisse eine Theorie dritter Ordnung für den räumlich beliebig gekrümmten Träger dar; sie können u. a. als Ausgangspunkt für die korrekte Formulierung von Stabilitätsproblemen dienen.

Streichung der nichtlinearen Terme in den Verzerrungen, Schnittgrößen und Trägheitsgliedern ergibt eine Theorie zweiter Ordnung dieses Trägers. Zusätzliche Streichung der aus der Formulierung der Gleichgewichtsaussagen am verformten System herrührenden Terme liefert Ergebnisse, die mit denen von [3] identisch sind, wenn dort die Verwölbungsanteile vernachlässigt werden. Sie stellen die lineare Theorie für den räumlich beliebig gekrümmten Träger dar. Es zeigt sich, daß die in [4] erhaltenen Ergebnisse für die Dehnung und damit für die Schnittgrößen inkorrekt sind; insbesondere kann bei linearen Ausgangsgleichungen kein sekundäres Torsionsmoment auftreten. Setzt man $\sqrt{g_0} \approx 1$, so ergeben sich für Schnittgrößen und Trägheitsterme wesentliche Vereinfachungen und man erhält eine Theorie dritter bzw. zweiter Ordnung für den schwach gekrümmten Träger, die bei Linearisierung in bekannte Ergebnisse, vergl. z. B. [5], übergeht.

Literatur

1 SCHROEDER, F. H., Allgemeine Stabtheorie des dünnwandigen, räumlich vorgekrümmten und verwundenen Trägers mit großen Verformungen, Ing.-Arch. **39**, 87–103 (1970).

2 BECKER, O., Der räumlich gekrümmte Träger bei endlichen Deformationen, Wiss. Berichte der Ing.-Hochschule Zittau, IHZ-M-79-214 (hochschulinterner Manuskriptdruck).

3 ARONSON, A. J., Ob odnom variante postroenija obobščennoj teorii steržnej, Sbornik: Dinamika i pročnost uprugich i gidrouprugich sistem, Moskau 1975, S. 99–114.

4 KÄMMEL, G., Zur Theorie des räumlich gekrümmten Stabes, Ing.-Archiv **27**, 255–267 (1959).

5 WAHL, F., Untersuchungen zur Kinetik räumlich gekrümmter Stäbe mit Anwendungen auf zylindrische Schraubenfedern, Dissertation, Techn. Hochschule Magdeburg 1970.

Eingereicht am 18. 9. 1978, in revidierter Fassung am 12. 7. 79

Anschrift: Dr.-Ing. OSWALD BECKER, Ingenieurhochschule Zittau, Abt. Mathematik, DDR-88 Zittau, Theodor-Körner-Allee 16, DDR

ZAMM 60, 118–120 (1980)

R. S. ALWAR / YOGENDRA NATH / B. SEKHAR REDDI

Axisymmetric Dynamic Buckling of Shallow Spherical Shells

This investigation is concerned with the dynamic axisymmetric buckling of shallow spherical shells under instantaneously applied step loading. Most of the investigations carried out till now employ numerical techniques to determine the dynamic buckling loads. Although several results are available in connection with this problem, there is no consistent agreement among the various results. HUANG [1] considered the problem of a spherical shell subjected to impulse loading and instantaneously applied step load. In his work nonlinear governing differential equations were solved by Finite Difference method with an iterative procedure. STEPHEN and FULTON [2] have solved the problem of partially loaded spherical cap by Finite Difference technique. STRICKLIN et al. [3] have used the general programme which was developed based on matrix displacement method to solve the problem of nonlinear elastic behaviour of shells of revolution subjected to arbitrary loading. In their work the nonlinear equilibrium equations are linearized by separating the linear and nonlinear portions of the strain energy and using the curved element. In 1973 BALL and BURT [4] used the Finite Difference scheme by reducing two fourth order partial differential equations to four second order partial differential equations and treating the nonlinear terms in the equations as pseudo loads. In 1974 SMITSES [6, 7] reviewed the problem of dynamic snapthrough buckling of shallow spherical caps and presented the MPCL and MGCL results based on energy methods. In the present investigation an alternative analytical procedure has been used to solve this problem. This procedure consists of applying the rapidly converging CHEBYSHEV series in the radial direction and the HOUBOLT [10] integration scheme in the timewise direction. The advantage of using CHEBYSHEV series in the accurate results can be obtained with as low as five terms in the series expansion. The governing nonlinear differential equations are linearized by using TAYLOR's expansion technique and subsequently solution is obtained by using CHEBYSHEV series. This procedure and its effectiveness has been demonstrated earlier by the authors [8, 9] in its application to nonlinear transient response of plates.

The governing differential equations for the nonlinear dynamic behaviour of a spherical cap in nondimensional form are as follows.

$$\varrho^3\nabla^4 W - 12(a/h)^2\,\varrho^2\left\{\frac{d}{d\varrho}\left(\psi\,\frac{dW}{d\varrho}\right) + K(h/a)\,\frac{d}{d\varrho}(\varrho\psi)\right\} = = \varrho^3\left(P - \frac{d^2W}{d\tau^2}\right) \tag{1}$$

$$\varrho^2(\nabla^2 - 1/\varrho^2)\,\psi + \frac{(1-\nu^2)}{2}\left[\varrho\left(\frac{dW}{d\varrho}\right)^2 + 2K(h/a)\,\varrho^2\left(\frac{dW}{d\varrho}\right)\right] = 0 \tag{2}$$

where $\nabla^2 = \frac{d^2}{d\varrho^2} + \frac{1}{\varrho}\frac{d}{d\varrho}$; a = base radius of the shell; t, h = thickness of the shell; K = nondimensional curvature of shell; τ = nondimensional time.

The solution of these equations can be expressed in terms of CHEBYSHEV series:

$$W = \sum_{r=0}^{N} \overline{W}_r T_r(\varrho)\,, \tag{3}$$

$$\psi = \sum_{r=0}^{N} \psi_r T_r(\varrho)\,, \tag{4}$$

where $\overline{W}_r$ and ψ_r $(r = 1 \dots N)$ are the CHEBYSHEV coefficients.

After linearizing equations (1) and (2) by TAYLOR's expansion [9] and substituting the dependent variables and their derivatives in CHEBYSHEV series and using HOUBOLT scheme [10] in the timewise direction, the nonlinear dynamic response of the shell can be obtained.

The criteria used for predicting the dynamic buckling loads is the same as suggested by BUDIANSKY and ROTHE [5]. The average deflection $\bar{\varrho}$ is plotted versus time and the load at which the sudden jump in the average deflection and simultaneously the reversal of the first maximum occurs is assumed to be dynamic buckling load. The parameter $\bar{\varrho}$ is defined as

$$\bar{\varrho} = \frac{2}{\lambda^2}\int_0^{\lambda} r w(r,\tau)\,dr \tag{5}$$

where $\lambda = 2[3(1-\nu^2)]^{1/4}\,(H/t)^{1/2}$ and w is nondimensional vertical displacement.

The plots in Fig. 1 and Fig. 2 show typical large amplitude shell responses for geometric parameters $\lambda = 5$ and $\lambda = 8$ from which the critical loads are predicted. The results obtained from the present analysis are compared with the existing results in Table 1. Fig. 3 illustrates the geometry of the shell and the variation of P_{cr} with λ. There is good agreement between the results of the present analysis and results given by BALL and BURT [4], for all values of λ except for $\lambda = 8$ and 9. For $\lambda = 8$, 9 there is poor agreement between the two results. This may be due to the illconditioned nature of the equations in the region $\lambda = 8$ and 9.

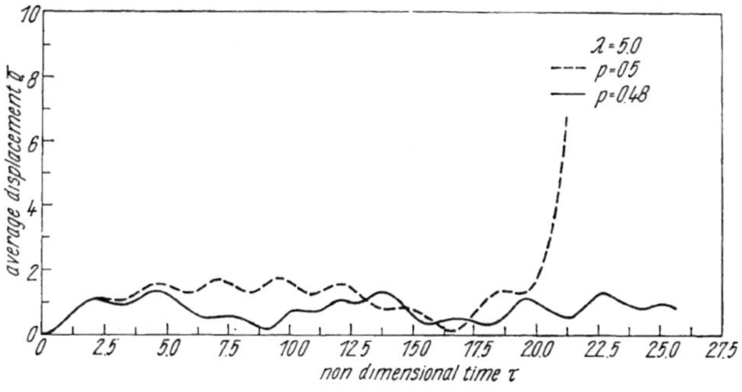

Fig. 1. Dynamic response

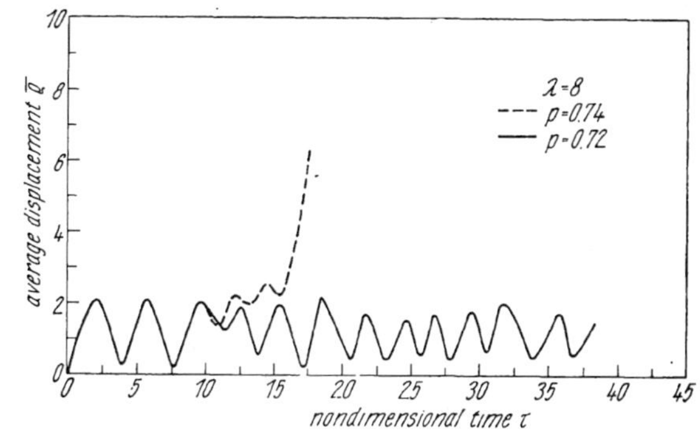

Fig. 2. Dynamic response

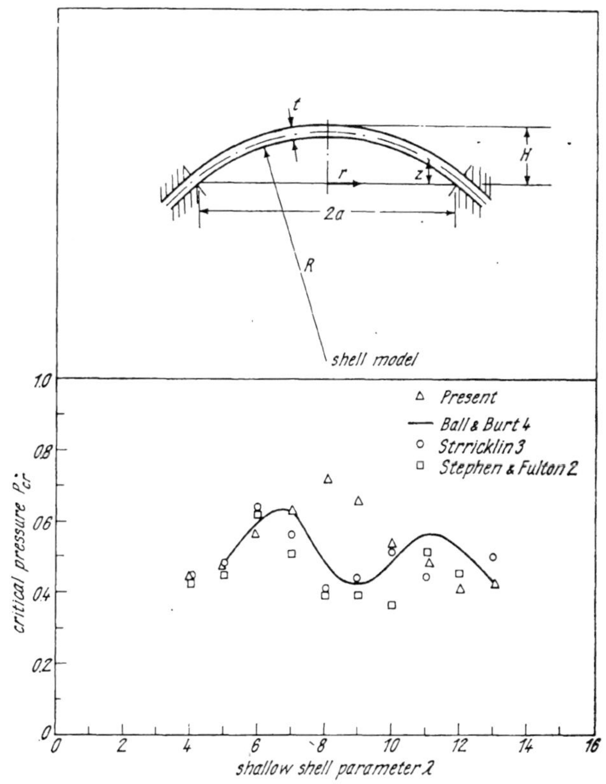

Fig. 3. Critical pressure vs shallow shell parameter

Table 1

	4	5	6	7	8	9	10	11	12	13
Authors	0.44	0.48	0.58	0.63	0.72	0.65	0.53	0.50	0.41	0.41
BALL and BURT	–	0.48	0.61	0.62	0.48	0.43	0.50	0.575	0.52	0.41
STEPHENS and FULTON	0.43	0.45	0.62	0.51	0.40	0.39	0.37	0.51	0.45	–
HUANG	0.45	0.49	–	0.56	0.44	0.39	0.42	0.50	–	0.42
STRICKLIN	0.44	0.48	0.64	0.56	0.47	0.41	0.43	0.54	0.44	0.50
SMITSES MPCL	0.42	0.4	0.38	0.4	0.45					
SMITSES MGCL	0.57	0.62	1.03	1.11	1.15					

In order to eliminate the influence of the magnitude of the time step on the buckling loads calculations were done for three values of $\Delta\tau$ viz. $\Delta\tau = 0.0005$, $\Delta\tau = 0.001$ and $\Delta\tau = 0.002$ and it was found that consistent results could be obtained for $\Delta\tau = 0.001$. Also the dynamic response study was conducted over a fairly long period of time in order to ensure that the dynamic loads obtained were not in the higher range. For all geometric parameters the response studies were conducted up to $\tau = 120$ as was done in Ref. [4]. Seven terms were used in the CHEBYSHEV series and it can be seen from Fig. 4 that there is good convergence even for five terms in series expansion. In the foregoing investigation the efficacy of the rapidly converging CHEBYSHEV series in solving the dynamic buckling problem has been demonstrated. This method can be extended to dynamic snapthrough buckling of orthotropic spherical caps and annular spherical shells and it is presently in progress. These are the problems currently under investigation by the authors.

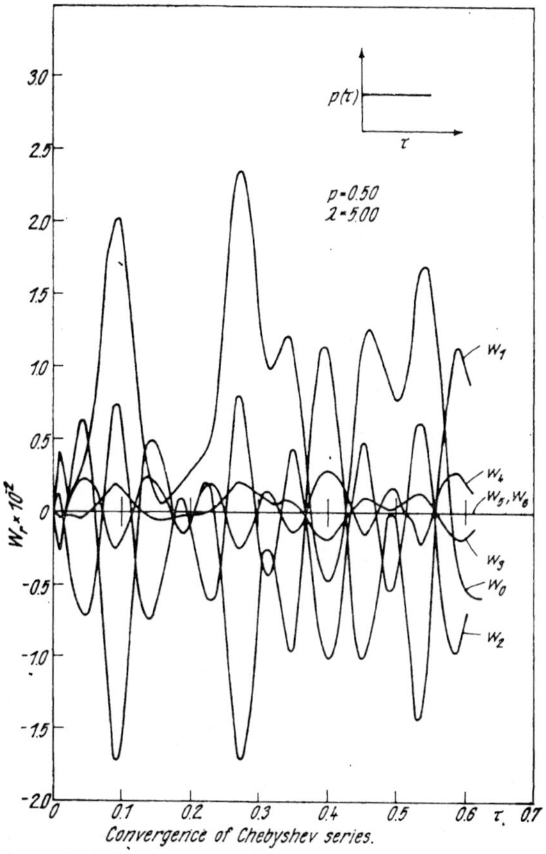

Fig. 4. Convergence of CHEBYSHEV series

Acknowledgement

The authors would like to thank the C.S.I.R., New Delhi for the financial assistance received from them in connection with this research work.

References

1 HUANG, N. C., Axisymmetric Dynamic Snapthrough of Elastic Clamped Shallow Spherical Shells, AIAA Journal, 7, pp. 215–220 (Feb. 1969).
2 STEPHENS, W. B. and FULTON, R. E., Axisymmetric Static and Dynamic Buckling of Spherical Caps due to Centrally Distributed Pressures, AIAA Journal, 7, 1969, pp. 2120–2126.
3 STRICKLIN, J. A. and MARTINEZ, J. E., Dynamic Buckling of Clamped Spherical Caps Under Pressure Loadings, AIAA Journal, 7, 1212–1213 (June 1969).
4 BALL, R. E. and BURT, J. A., Dynamic Buckling of Shallow Spherical Shells, Journal of Applied Mechanics, **40**, 11, pp. 411–416 (June 1973).
5 BUDIANSKY, B. and ROTH, R. S., Axisymmetric Dynamic Buckling of Clamped Shallow Spherical Shells, NASA TND 1510, 1962, pp. 597–606.
6 SMITSES, G. J., Axisymmetric Dynamic Snapthrough of Elastic Clamped Shallow Spherical Shell, AIAA Journal, Vol. **5**, No. 5, 1967, pp. 1019–1021.
7 SMITSES, G. J., On the Buckling of Shallow Spherical Shells, Journal of Applied Mechanics, ASME, March 1974, pp. 299–300.
8 ALWAR, R. S. and NATH, YOGENDRA, Application of CHEBYSHEV Polynomials to the Nonlinear Analysis of Circular Plates, Int. J. Mech. Sci., Vol. **18**, pp. 589–595, 1976.
9 ALWAR, R. S. and NATH, YOGENDRA, Nonlinear Dynamic Response of Circular Plates Subjected to Transient Loads, Journal of the Franklin Institute, Vol. **6**, pp. 527, 1977.
10 HOUBOLT, A Recurrence Matrix Solution for the Dynamic Response of Elastic Aircraft, J. Aero. Sci., Vol. **17**, No. 8, September 1950, pp. 540–550.

Eingereicht am 1. 2. 1979

Anschrift: Prof. R. S. ALWAR, Department of Applied Mechanics, I. I. T., Madras 600036, India

BUCHBESPRECHUNGEN

Schwetlick, H., Numerische Lösung nichtlinearer Gleichungen. Berlin VEB Deutscher Verlag der Wissenschaften 1979. 346 S., 21 Abb., M 68,–.

The aim of the book is to present in German a survey of the most important algorithms for the solution of finite dimensional nonlinear problems. Only algorithms derived from NEWTON's method or the gradient-method are considered. This excludes, for instance, the general class of fixed-point methods which received considerable attention in recent years.

The book is addressed to numerical mathematicians and mathematically oriented engineers and scientists. Without doubt it may also serve as a text for graduate students. All results are proved in detail and the important algorithms are presented in an informal language. Each section ends with a set of exercises and there are numerous bibliographic notes and remarks.

The range of material is best summarized by a listing of the chapter headings: 1) Problem formulation, 2) Mathematical background, 3) Existence problems and characterization of solutions, 4) General remarks on iterative methods, 5) Locally superlinearly convergent methods for finding zeros, 6) General theory of descent methods, 7) Descent methods for the minimization of uniformly convex functions, 8) Descent methods for the minimization of non-convex functions, 9) Survey of derivative-free methods for the minimization on nonlinear functions, 10) Methods for the minimization of a sum of squares, 11) Continuation methods for the solution of nonlinear equations.

The overall presentation in form and method of approach appears to have been strongly influenced by the book of J. ORTEGA and the reviewer ["Iterative Methods for Nonlinear Equations in Several Variables", Academic Press, Inc 1970]. But, of course, this books covers more discrete NEWTON methods, included are, in particular, some of the newer discrete NEWTON methods, certain globally convergent minimization algorithms, and the author's own work on continuation techniques. With nearly a third of the pages, the minimization methods clearly assume a central place in the book. In the selection of the newer material, it is not surprising that more of the Eastern-European than, say, the American literature has been taken into account. But this certainly does not detract from the value of the book as a very worthy addition to the literature in this burgeoning field.

Pittsburgh, PA W. RHEINBOLDT

Sabatier, P. C. (ed.), Applied Inverse Problems. Lectures Presented at the RCP 264 "Etude interdisciplinaire des problèmes inverses". Lecture Notes in Physics 85. Berlin-Heidelberg-New York, Springer-Verlag 1978. V, 425 S., 37 Abb., 13 Tab., DM 39,–. US $ 21,50. ISBN 3-540-09094-0

Since 1972 the editor has organized conferences on inverse problems in Montpellier every year. In these Proceedings the lectures of the conference in 1977 are published. The book is divided into two parts. Part I: Fifteen review lectures on applied inverse problems. Part II: Five lectures on special applications and one theoretical lecture on solutions of inverse problems. We learn from the introductory article the following remarks: Inverse problems began in physics as soon as physicists tried to fit experimental results into a theoretical model. But the first theoretical models were given by such simple rules that there was hardly a problem. ... In this way, which is a general fitting technique, one reduces the set of theoretical parameters to a subset depending on very few parameters, which are determined by trial and error, using the direct problem. Although these techniques always ignore intrinsic ambiguities, and most error questions, it is only when they fail that physicists remember that studying inverse problems is the only complete way of analyzing experimental results. ... One can find further papers from a conference which took place in 1979, in "Inverse and Improperly Posed Problems in Differential Equations", Akademie-Verlag Berlin.

Berlin G. ANGER

Anger, G. (ed.), Inverse and Improperly Posed Problems in Differential Equations, Proceedings of the Conference on Mathematical and Numerical Methods Held in Halle/Saale (GDR), May 29 – June 2, 1979. Mathematical Research / Mathematische Forschung 1. Berlin, Akademie-Verlag 1979. 296 S., 37,00 M. Best.-Nr. 7627441 (2182/1)

Inverse and incorrectly posed problems have many important applications in various applied sciences and in technology. Problems of this type usually occur when one has to determine unknown parameters or functions which govern a physical process, by means of data obtained by an observation of the system under consideration. Typical and classical examples arise in the context of systems governed by the heat equation where it is often desired to estimate the unknown initial temperature distribution of a body (the earth f. i.) by measurements at later times. — The Proceedings of the conference at Halle reflects the fact that there are important connections of the theory of inverse and ill-posed problems to other fields of applied mathematics, for example to statistics, control theory, potential theory, and general operator theory. The volume contains theoretically as well as numerically orientated contributions and also some interesting papers on fields of applications of inverse problems (seismics, spectroscopy, vibrating systems, fluid dynamics, and others). There are 28 articles (25 of them in English language) written by experts from many different countries: G. ANGER and R. CZERNER (GDR), V. BARCILON (USA) and G. TURCHETTI (Italy), J. R. CANNON and R. E. EWING (USA), G. CHAVENT (France), D. COLTON (USA), J. ELDÉN (Sweden), V. FRIEDRICH (GDR), I. GALLIGANI (Italy), K. GLASHOFF (FRG), M. GOEBEL (GDR), A. GÖPFERT (GDR), R. GORENFLO (Berlin-West), V. M. ISAKOV (USSR), R. KLUGE (GDR), R. J. KNOPS and L. E. PAYNE (USA), R. KÜHNAU (GDR), M. M. LAVRENTIEV (USSR), B. MICHEL (GDR), J. NEDOMA (ČSSR), R. NÜRNBERG (GDR), J. PEIL (GDR), V. G. ROMANOV (USSR), P. C. SABATIER (France), T. I. SEIDMAN (USA), G. STOYAN (GDR), M. TASCHE (GDR), D. ZIDAROV (Bulgaria), V. A. MOROZOV (USSR). — The conference has shown again very impressingly the importance and applicability of this relatively new field of mathematics and will certainly have positive influence on the development of new methods. The papers collected in the Proceedings are of interest for anyone who is working with inverse or incorrectly posed problems in differential equations.

Hamburg K. GLASHOFF

Eisenack, G. / Fenske, C., Fixpunkttheorie. Mannheim-Wien-Zürich, B. I.-Wissenschaftsverlag 1978. 258 S.,

Die Fixpunkttheorie behandelt die Frage, unter welchen Bedingungen eine stetige Abbildung eines topologischen Raumes in sich Fixpunkte besitzt. Dieses Problem ist seiner Natur nach eine topologische Aufgabe. Daher findet man die ersten Darstellungen der Fixpunkttheorie in Werken der Topologie (vgl. die „Topologie" (1935) von P. ALEXANDROFF und H. HOPF, oder die „Algebraic topology" (1942) von S. LEFSCHETZ). Das Eindringen funktionalanalytischer Methoden in die Fixpunkttheorie sowie die zahlreichen Anwendungsmöglichkeiten, die die Fixpunkttheorie in der Analysis und in der numerischen Mathematik gefunden hat, führten in neuerer Zeit zur Entstehung eines Bildes der Fixpunkttheorie, das sich von demjenigen vor 30 Jahren wesentlich unterscheidet. War die Fixpunkttheorie früher nur ein Teilgebiet der Topologie, so kann man sie heute dank ihres Umfangs und Inhalts ohne weiteres als selbständiges Gebiet der modernen Mathematik ansehen, das zwischen Topologie und Funktionalanalysis liegt. Zweck des vorliegenden Buches ist es, die wichtigsten Problemstellungen, Methoden und Ergebnisse dieses Gebiets darzustellen. Es ist vor allem für Mathematiker geschrieben, die sich eine Übersicht über den heutigen Stand der Fixpunkttheorie verschaffen oder sich forschungsmäßig den noch vielen offenen Problemen auf dem Gebiet der Fixpunkttheorie zuwenden wollen.

Das Buch ist in neun Kapitel gegliedert. Im ersten Kapitel wird das bekannteste Ergebnis der Fixpunkttheorie vorgestellt, das Prinzip der kontrahierenden Abbildungen. Darauf folgt im zweiten Kapitel die Behandlung von Fixpunkträumen. Das sind topologische Räume, welche die Eigenschaft haben, daß jede stetige Abbildung dieser Räume in sich mindestens einen Fixpunkt besitzt. Es wird gezeigt, daß jede abgeschlossene Vollkugel im R^n und jede kompakte konvexe Menge eines lokalkonvexen Raumes ein Fixpunktraum ist (BROUWERscher Fixpunktsatz, Fixpunktsatz von SCHAUDER-TYCHONOFF). Das dritte Kapitel ist dem LEFSCHETZschen Fixpunktsatz gewidmet, der mit Hilfsmitteln aus der algebraischen Topologie bewiesen wird. In den folgenden beiden Kapiteln wird mit analytischen Mitteln ausführlich der Abbildungsgrad für differenzierbare und für stetige Abbildungen beschrieben. Nach einer kurzen Einführung in die Verzweigungstheorie im sechsten Kapitel folgt dann im siebenten Kapitel die axiomatische Einführung des Fixpunktindex. Existenz und Eindeutigkeit des Fixpunktindex werden vermittels des Abbildungsgrades bewiesen. Im darauffolgenden Kapitel wird der Fixpunktindex benutzt, um asymptotische Fixpunktsätze zu gewinnen. Das letzte, mit Varia überschriftete Kapitel behandelt Fixpunkte homotoper Abbildungen, Projektionsmethoden und Fixpunkte mengenwertiger Abbildungen. Das Buch schließt mit einem Anhang, in dem Sätze über FREDHOLM-Operatoren und absolute Umgebungsretrakte sowie Ergänzungen zur Homologietheorie zusammengefaßt sind. Es muß noch erwähnt werden, daß nach jedem Kapitel Ergänzungen und Bemerkungen folgen, in denen auf die benutzten Quellen hingewiesen wird und in denen weiterführende Angaben gemacht werden.

Das Buch kann allen, die sich über die Fixpunkttheorie informieren wollen, bestens empfohlen werden. Es ist den Verfassern gelungen, auf gedrängtem Raum eine Fülle von Ergebnissen in einer klaren und mathematisch einwandfreien Darstellung zu bringen. Der Wert des Buches wird noch erhöht durch vielfältige, in den Text eingestreute Beispiele, welche die Anwendbarkeit der gebrachten Sätze in verschiedenen Teilgebieten der Mathematik illustrieren.

Cluj-Napoca W. W. BRECKNER

Medeiros, L. A. J. / de la Penha, G. M. (eds.), Contemporary Developments in Continuum Mechanics and Partial Differential Equations. Proc. Internat. Sympos., Rio de Janeiro, Brazil, August 1977. North-Holland Mathematics Studies 30. Amsterdam-New York, North-Holland Publ. Cy. 1978. VIII, 614 S., US $ 53,25. Dfl. 120,–. ISBN 0-444-85166-6

Dieser Band vereinigt praktisch alle Vorträge, die auf einem internationalen Symposium über aktuelle Forschungen auf den Gebieten Kontinuumsmechnaik und partielle Differentialgleichungen gehalten wurden, das im August 1977 in Rio de Janeiro stattfand. Dieses Symposium hatte das Ziel, Mathematiker und Wissenschaftler anderer Disziplinen mit Hauptlinien der modernen Forschung auf diesen Gebieten bekannt zu machen. Zudem wollte es zu einer Bestandsaufnahme der Ergebnisse beitragen, die die in den letzten Jahren zu registrierende intensive Wechselwirkung zwischen beiden Gebieten gezeigt hat, und einen Ausblick auf künftige Entwicklungen geben. Die Qualität der folgenden in diesem Band enthaltenen Arbeiten zeigt, daß das Symposium seine Ziele erreicht hat:

S. S. ANTMAN, A family of semi-inverse problems of nonlinear elasticity; G. S. S. ÁVILA and D. G. COSTA, Asymptotic properties of general symmetric hyperbolic systems; J. B. BAILLON and J. M. CHADAM, The CHAUCHY problem for the coupled SCHROEDINGER-KLEIN-GORDON equations; T. B. BENJAMIN, Applications of generic bifurcation theory in fluid mechanics; H. BREZIS, The HAMILTON-JACOBI-BELLMAN equation for two operators via variational inequalities; H. BREZIS, Nonlinear problems related to the THOMAS-FERMI equation; F. CARDOSO and J. JOUNIE, Global solvability and hypoellipticity of abstract complexes and equations; MICHAEL O'CARROLL, On the inverse scattering problem for linear evolution equations; CLAUDE DO, Nonlinear bifurcation problem and buckling of an elastic plate subjected to unilateral conditions in its plane; B. D. COLEMAN, On the thermodynamics of non-classical systems; R. L. FOSDICK, On an inequality in thermodynamic stability; R. GLOWINSKI and O. PIRONNEAU, Least square solution of nonlinear problems in fluid dynamics; P. PODIO GUIDUGLI, Elastic bodies in a SIGNORINI-type environment; M. E. GURTIN, On the nonlinear theory of elasticity; D. D. JOSEPH, Constitutive equations and free surfaces; J. L. LIONS, On some questions in boundary value problems of mathematical physics; R. C. MACCAMY, Memory effects in one-dimension problems of continuum mechanics; WALTER NOLL, A general framework for problems in the statics of finite elasticity; P. NOWOSAD, Elliptic metrics on LORENTZ manifolds; S. OSHER, Boundary value problems for equations of mixed type; J. P. PUEL, A free boundary, nonlinear eigenvalue problem; JAMES SERRIN, The concepts of thermodynamics; W. A. STRAUSS, The nonlinear SCHRÖDINGER equation; G. SVETLICHNY and P. OTTERSON, Derivative dependent infinitesimal deformations of differentiable functions; L. TARTAR, Nonlinear constitutive relations and homogenization; R. TEMAN, Qualitative properties of NAVIER-STOKES equations; G. TRUESDELL, Some challenges offered to analysis by rational thermodynamics; W. O. WILLIAMS, Some simplified equations from the theory of mixtures.

Berlin H. GAJEWSKI

Knobloch, H. W. / Kappel, F., Gewöhnliche Differentialgleichungen. Stuttgart, B. G. Teubner 1974. 322 S, DM 48,–.

Das vorliegende Lehrbuch über gewöhnliche Differentialgleichungen zeichnet sich durch folgende Eigenschaften aus:

1. Es stellt einen Leitfaden im eigentlichen Sinne des Wortes dar, d. h. es werden die wesentlichen Resultate der Theorie linearer und nichtlinearer gewöhnlicher Differentialgleichungen unter plausiblen Voraussetzungen (z. B. LIPSCHITZ-Stetigkeit der rechten Seite) in überschaubarer Weise dargestellt, ohne dabei „im Vorfeld gerade jener Probleme stehenzubleiben, derentwillen man zumeist ein Buch über Differentialgleichungen zur Hand nimmt". Gleichzeitig liegt der Darlegung eine einheitliche Konzeption zugrunde, so daß auch verschiedenartige und scheinbar unzusammenhängende Gebiete als Teile eines einheitlichen Ganzen erkannt werden.

2. Neben den traditionellen Gegenständen der Lehrbücher über gewöhnliche Differentialgleichungen (elementare Integrationsmethoden, lineare Differentialgleichungen, Existenz- und Eindeutigkeitsfragen) stellt die Theorie nichtlinearer Systeme den eigentlichen Schwerpunkt des Buches dar. Dabei bleiben die Verfasser nicht bei der POINCARÉ-BENDIXSONschen Theorie ebener autonomer Systeme und der direkten Methode von LJAPUNOV zur Behandlung von Stabilitätsproblemen stehen, sondern setzen mit einem neuartigen Zugang zur Theorie der Integralmannigfaltigkeiten, der sich organisch aus der Darlegung der allgemeinen Theorie gewöhnlicher Differentialgleichungen ergibt, und einer neuen Variante zum Beweis der notwendigen Bedingungen (PONTRJAGINsches Maximumprinzip) für die optimale Lösung eines Steuerungsproblems gewöhnlicher Differentialgleichungen zwei markante Höhepunkte.

Das vorliegende Lehrbuch, zu dessen Studium die Kenntnisse aus den mathematischen Grundvorlesungen des ersten Studienjahres (Analysis I, II, Lineare Algebra) benötigt werden, kann man neben den Mathematikern auch allen Naturwissenschaftlern und Ingenieuren für Regelungstechnik empfehlen, die Systeme gewöhnlicher Differentialgleichungen als mathematische Modelle verschiedenartiger Prozesse verwenden. Im einzelnen enthält das Buch folgende Kapitel: 1. Elementare Integrationsmethoden. 2. Lineare Differentialgleichungen. 3. Allgemeine Theorie nichtlinearer Differentialgleichungen. 4. Ebene autonome Systeme. 5. Linearisierung. 6. Optimierung. 7. Konvexe Kegel im R^n. 8. Lösungen zu ausgewählten Aufgaben.

Berlin K. R. SCHNEIDER

Stoer, J. / Bulirsch, R., Einführung in die Numerische Mathematik II. unter Berücksichtigung von Vorlesungen von **F. L. Bauer.** Heidelberger TB 114. 2., neubearb. Auflage. Berlin-Heidelberg-New York, Springer-Verlag 1978. XI, 308 S., 22 Abb., DM 19,80, US $ 9.90.

Der in dieser Zeitschrift bereits besprochenen, 1973 erschienenen ersten Auflage der „Einführung in die Numerische Mathematik, Teil II" hat der Verlag nun die auf Grund der außerordentlich guten Resonanz zu erwartende zweite Auflage folgen lassen. Die Autoren haben dabei die Mühe einer Neubearbeitung nicht gescheut und neben kleineren Änderungen einige neue Gebiete mit eingearbeitet. So ist im Kapitel über Eigenwertprobleme die theoretisch wie praktisch bedeutsame Singulärwertzerlegung einer Matrix und das GOLUB-REINSCH-Verfahren zu deren Berechnung aufgenommen worden. Allerdings wurde die Möglichkeit, auf dieser Grundlage in einfacher Weise Probleme der Korrektheit oder Inkorrektheit von linearen Gleichungssystemen bzw. linearen Quadratmittelproblemen zu diskutieren und etwa regularisierte Verfahren zur stabilen Lösungsberechnung abzuleiten, nicht genutzt, wie generell der Bezug auf Kapitel 4 des ersten Teiles lediglich in der Bemerkung besteht, daß aus der Singulärwertzerlegung die Pseudoinverse berechnet werden kann.

Im Kapitel über gewöhnliche Differentialgleichungen sind praktisch wichtige Methoden zur automatischen Schrittweitensteuerung bei RUNGE-KUTTA- und Mehrschrittverfahren sowie ein Abschnitt über steife Systeme aufgenommen worden, der allerdings nur Überblickscharakter und nicht die Ausführlichkeit der übernommenen Abschnitte besitzt. Ferner wurde als Beispiel für die in den letzten Jahren entwickelten „schnellen" Algorithmen das Verfahren von BUNEMAN in die Darstellung aufgenommen.

Trotz der Probleme, die mit der nachträglichen Aufnahme neuer Gebiete in bewährte Texte verbunden sind und die auch hier die Ausgewogenheit der ursprünglichen Fassung etwas gemindert haben, hat sich der ohnehin hohe Wert des Buches als eine Maßstäbe setzende Einführung in die Numerische Mathematik durch die vorgenommene Modernisierung weiter erhöht.

Dresden H. SCHWETLICK

Böhmer, K., Spline-Funktionen. Theorie und Anwendungen. Stuttgart, B. G. Teubner 1974. 340 S., 14 Abb., DM 24,80.

In dem Buch wird eine mathematisch fundierte, anspruchsvolle Einführung in die Theorie der Spline-Funktionen einer reellen Variablen gegeben. Ausgehend von den Schwierigkeiten bei der Interpolation durch Polynome wird die Entwicklung der Spline-Funktionen behandelt. In moderner Darstellung werden unter Verwendung funktional-analytischer Mittel einerseits Darstellungen von Spline-Funktionen gebracht, die wichtig sind für eine übersichtliche Beweisführung und die Theorie, und andererseits die für die numerischen Anwendungen bedeutsamen Polynomsplines, für die der nur anwendungsinteressierte Leser eine von den übrigen theoretischen Kapiteln unabhängige Einführung findet. Durch ein gesondertes Kapitel wird auch der Zugang zu den theoretischen Teilen des Buches unterstützt.

In den numerisch orientierten Kapiteln wird die Behandlung von Anfangs-, Rand- und Eigenwertaufgaben mit Splines behandelt. Die Spline-Verfahren für Anfangswertaufgaben führen einerseits auf bekannte implizite Mehrschrittverfahren, andererseits ergeben sich bei der Verwendung von HERMITE-Splines A-stabile Verfahren beliebig hoher Ordnung. Die Anwendung der Splines auf Randwertaufgaben erfolgt im Zusammenhang mit dem Verfahren von RAYLEIGH-RITZ-GALERKIN und der Kollokationsmethode. Theoretische und experimentelle Vergleiche mit anderen numerischen Verfahren fehlen nicht. In einem Anhang befinden sich ALGOL-Programme.

Berlin G. HEBERMEHL

Reinschke, K. / Schwarz, P., Verfahren zur rechnergestützten Analyse linearer Netzwerke. Berlin, Akademie-Verlag 1976. XI, 324 S., 156 Abb., M 42,—. Best-Nr. 762092 8 (6251)

Die enorme Fortentwicklung der Theorie elektrischer Netze ist vor allem auch den Fortschritten und der großen Verbreitung des Digitalrechners zu verdanken. So wird auch heute bei Verfahren zur Berechnung elektrischer Netze nicht mehr besonders erwähnt, daß die Berechnung mit einem Digitalrechner zu erfolgen hat. Besonders beliebt und wirkungsvoll sind heute Netzberechnungen im Dialogbetrieb am Bildschirm geworden. Diese als „rechnergestützt" bezeichneten Verfahren erlauben eine optimale Ausnutzung vorhandener Programme. Hier müssen auch nicht alle notwendigen Entscheidungen bereits im Programm festgelegt sein, zumal sich auch manche Intuitionen nur schwer formalisieren lassen.

Die Autoren dieses Buches haben es dankenswerterweise übernommen, die fast unübersehbare Flut von Arbeiten in einem gut gelungenen Lehrbuch zusammenzufassen. Dabei beschränken sich die Verfasser auf die reinen Analysemethoden, insbesondere soweit sie in der Nachrichtentechnik Anwendung finden. Dort ist dann auch die Kenntnis der Analyse elektrischer Netze Voraussetzung zur Netzwerksynthese. Freilich können viele Verfahren der nachrichtentechnischen Netzwerkanalyse auch in der elektrischen Energietechnik eingesetzt werden.

Eine gut fundierte Theorie elektrischer Netze muß sich auch mit den Grundlagen der Graphentheorie auseinandersetzen. Dies geschieht nach einem einführenden ersten Kapitel bereits im zweiten Kapitel. Bei der Definition der grundlegenden Schaltelemente werden gelegentlich auch nichtelektrische Schaltelemente zum Vergleich herangezogen. In diesem zweiten Kapitel werden dann auch die Gleichungssysteme zur Analyse linearer Netze mit Ein- und Mehrtorelementen aufgestellt. Die Analyse im Frequenzbereich wird im dritten Kapitel dargestellt, wo außer numerischen Methoden auch symbolische Methoden näher erläutert werden. Zur Untersuchung von Filtern und Weichen ist es häufig wichtig, zur Weiterverarbeitung formelmäßige Ausdrücke zur Hand zu haben. Der Rechenaufwand durch den Digitalrechner ist dabei freilich größer als bei rein numerischen Verfahren, dafür ist die Aussagekräftigkeit der symbolischen Ergebnisse auch vielfältiger. Hier sollen nur Stichworte erwähnt werden: topologische Methoden, Strukturzahlen, WANG-Algebra, Methode der verallgemeinerten Zahlen. Die symbolischen Methoden stehen in Konkurrenz zu dem Verfahren der punktweisen Frequenzanalyse. In einem weiteren Kapitel werden die wichtigsten Methoden der Analyse im Zeitbereich dargestellt. Vielleicht sollte hier auch noch auf Methoden zur Beseitigung unerwünschter Erscheinungen in den Lösungskurven, die durch das GIBBSsche Phänomen hervorgerufen werden können, eingegangen werden. Ein abschließendes Kapitel behandelt Toleranzanalyse, Rauschanalyse und Optimierung im Bereich der Synthese.

Das vorliegende didaktisch gut angelegte Buch kann nicht nur Studierenden, sondern auch Praktikern bestens empfohlen werden. Es füllt eine merkliche Lücke auf diesem Gebiet aus.

Uttenreuth H. EDELMANN

Koslow, B. A. / Uschakow, I. A., Handbuch zur Berechnung der Zuverlässigkeit in Elektronik und Automatentechnik. In deutscher Sprache herausgeg. und ergänzt v. **K. Reinschke.** Elektronisches Rechnen und Regeln Bd 25. Berlin, Akademie-Verlag 1978. XIX, 598 S., 158 Abb., 188 Tab., 78,— M. Best.-Nr. 7 623 467 (6300)

Die Herausgabe des Handbuches in deutscher Sprache ist sehr zu begrüßen: Die Thematik ist aktueller denn je, und ein vergleichbares Buch fehlte bisher. Es handelt sich um eine Übersetzung und wesentliche Ergänzung der 2. russischen Ausgabe von 1975. — Dieses Kompendium enthält 14 Kapitel: 1. Bezeichnungen und Begriffe der Zuverlässigkeit. Zuverlässigkeitskenngrößen. 2. Zuverlässigkeit eines Elementes, 3. Reservierung von Systemen ohne Erneuerung. 4. Reservierung mit Erneuerung. 5. Einige spezielle Klassen von Systemen mit Erneuerung und Reserve. 6. Seriensysteme im Sinne der Zuverlässigkeit. 7. Berücksichtigung des Einflusses von Kontroll- und Umschalteinrichtungen auf die Zuverlässigkeit. 8. Systeme mit komplizierter Struktur. 9. Abschätzung der Effektivität der Arbeitsweise von Systemen. 10. Aufgaben der optimalen Reservierung. 11. Aufgaben der optimalen Instandhaltung. 12. Optimale Fehlersuche. 13. Mathematisch-statistische Methoden zur Auswertung experimenteller Daten über die Zuverlässigkeit. 14. Zuverlässigkeit technischer Erzeugnisse in Abhängigkeit ihrer Umgebungs- und Einsatzbedingungen. Dazu kommen 8 Anhänge: 1. Grundbegriffe und kleiner Wissensspeicher der Wahrscheinlichkeitstheorie. 2. Grundbegriffe und kleiner Wissensspeicher der mathematischen Statistik. 3. Ströme (Punktprozesse). 4. Abriß der BOOLEschen Zuverlässigkeitstheorie von Systemen. 5. Systembeschreibung durch zufällige Prozesse mit endlich vielen Zuständen. 6. Heuristische Methode zur Abschätzung der Zuverlässigkeit von Systemen mit Erneuerung. 7. Einige Konstanten und Formeln, die bei der Lösung von Zuverlässigkeitsaufgaben häufig vorkommen. 8. Zahlentafeln. — Das Buch wendet sich an Benutzer, die schon eine gewisse, aber nicht unbedingt intime Bekanntschaft mit der Zuverlässigkeitstheorie gemacht haben und an deren Anwendungen interessiert sind. Solche Anwendungen hat es zwar bisher besonders „in Elektronik und Automatentechnik" gegeben, trotzdem sollte dieser Zusatz im Titel vielleicht gestrichen werden: Der Text macht kaum Gebrauch davon, wozu also Interessenten aus anderen Branchen unnötig verschrecken. Das Buch ist als Nachschlagewerk, nicht als Lehrbuch konzipiert. Dementsprechend wird der Stoff rezeptartig, häufig in Übersichten und Tabellen dargeboten: Definition der Grundbegriffe, Angabe von Berechnungsformeln, Anwendung der Berechnungsverfahren an Beispielen, Aufzählung wesentlicher Literatur; die Herleitung der Ergebnisse ist nicht Sache des Buches; ein Abriß der Theorie wird jedoch in den Anhängen gegeben. — Dieses Kompendium wird bei vielen Fragen der Zuverlässigkeitsarbeit ein wertvoller Ratgeber sein.

Berlin W. HEINRICH

Paul, W. J., Komplexitätstheorie. Stuttgart, B. G. Teubner 1978. 247 S., 25,80 DM.

Das Buch ist als Lehrbuch für Informatik- und Mathematikstudenten gedacht. Demgemäß setzt es nur wenige Grundkenntnisse dieser Gebiete voraus und enthält am Ende eines jeden

Abschnitts Übungsaufgaben. Für weitergehend Interessierte sind den Abschnitten Zusammenstellungen von z. Zt. offenen Problemen angefügt. Vom Inhalt her liegt der Schwerpunkt des Buches in erster Linie auf den verschiedenen Möglichkeiten der Präzisierungen des Begriffes „Kompliziertheit" und deren Eigenschaften; behandelt werden: Begriffe der Berechenbarkeit von Funktionen durch TURING-Maschinen (die im folgenden eine dominierende Rolle spielen) bzw. Rekursivität und ihre Äquivalenz, nicht entscheidbare Probleme (damit zusammenhängend die Unvollständigkeit der Arithmetik), verschiedene Kompliziertheitsmaße (bezüglich Zeit bzw. Speicher) und die (polynomiale) Äquivalenz der entsprechenden Begriffe für Register- bzw. TURING-Maschinen, die Rolle des Nichtdeterminismus, Hierarchien von Kompliziertheiten und vollständige Probleme in bezug auf verschiedene Kompliziertheitsklassen. Als letztes wird die axiomatische Kompliziertheitstheorie kurz dargestellt; Fragen der Kompliziertheit konkreter Algorithmen werden nicht behandelt. Die Darstellung des Stoffs ist gut lesbar, trotz erfrischend aufgelockertem Stil stets korrekt und in den Beweisen vollständig. Besonders wird deutlich herausgestellt, durch welche wenigen wesentlichen Techniken (Diagonalisierung und Reduktion) die meisten Resultate erzielt worden sind.

Berlin W. ISSEL

Wechler, W., The Concept of Fuzziness in Automata and Language Theory. Studien zur Algebra und ihre Anwendungen 5. Berlin, Akademie-Verlag 1978. 148 S., M 27,–.

After an introductory chapter on fuzzy sets and fuzzy logic, the book under review deals with R-fuzzy automata and grammars. (The basic notion is that of an R-fuzzy event: a mapping from Σ^* into a po-semiring R.) The main emphasis is on generative capacity and closure properties. Also interconnections with automata with multiplicities and various control structures, such as time-variant automata, are discussed. In view of the growing interest in fuzzy sets, the book under review is welcome and fills a gap in the existing literature. Apart from people working on fuzzy sets, this book should be of interest to people working on classical automata and language theory, as well as on formal power series in noncommuting variables. It is well written and also self-contained. Starting from the beginning, it gives all the definitions needed. Also some open problems are mentioned. The English spelling contains unusually many errors, which makes the reading sometimes uncomfortable.

Turku A. SALOMAA

Salomaa, A. K., Formale Sprachen. Übersetzt a. d. Engl. v. **E.-W. Dieterich.** Berlin-Heidelberg-New York, Springer-Verlag 1978. XI, 314 S., 18 Abb., DM 48,–, US $ 26,40. ISBN 3-540-09030-4

Das Buch gibt eine Einführung in die meisten der gegenwärtig wesentlichen formalen Sprachen. Es seien hier genannt:

- die klassischen Sprachen der CHOMSKY-Hierarchie: reguläre, kontextfreie, kontext-sensitive und Typ-0-Sprachen,
- theoretisch bzw. im Zusammenhang mit der Sprachanalyse interessante spezielle kontextfreie Sprachen wie lineare, LR(k)-, LL(k)-Sprachen und Sprachen von endlichem Index,
- Grammatiken mit einem gesteuerten Ableitungsmechanismus wie Matrix-, programmierte, zeitvariierende Grammatiken,
- nichtsequentielle Ersetzungsverfahren wie LINDENMAYER-Systeme und Scattered-Kontext-Grammatiken.

Die Theorie der formalen Sprachen wird dabei weitestgehend vom Standpunkt der Erzeugungsverfahren, insbesondere Grammatiken, betrachtet, aber auch die Beziehungen zur Automatentheorie, vereinheitlichende Prinzipien wie die Theorie abstrakter Familien von Sprachen, Fragen der algorithmischen Lösbarkeit von Problemen und Kompliziertheitsuntersuchungen werden diskutiert.

Diese Anlage des Buches ermöglicht eine Nutzung in zwei Richtungen. Zum einen kann es als Lehrbuch der formalen Sprachtheorie dienen. Diese Möglichkeit wird dadurch unterstützt, daß der Autor nur elementare Kenntnisse der Algebra und Logik voraussetzt, neben den formalen Definitionen auch verbale, die Grundidee beschreibende Erklärungen gibt, äußerst geschickt gewählte Beispiele zur Erläuterung der Begriffe verwendet und vielfach kurze Randbemerkungen macht, die die möglichen Schwierigkeiten eines Anfängers beseitigen sollen. In diesem Zusammenhang sei auch eine Lösung der Übungsaufgaben empfohlen. Zum anderen dringt der Verfasser jeweils bis zu einigen wesentlichen grundlegenden Ergebnissen der behandelten Theorien vor, so daß das Buch auch vom Spezialisten als Nachschlagewerk verwendet werden kann.

Ein weiterer Vorteil des Buches besteht darin, daß alle für die Theorie der Erzeugungsverfahren wichtigen Aussagen vollständig bewiesen werden (so z. B. auch der Satz von CHOMSKY-SCHÜTZENBERGER und die algorithmische Unlösbarkeit des POSTschen Korrespondenzproblems, deren Beweise in anderen Büchern oft nur skizziert sind). Die Übersetzung kann als gelungen angesehen werden (zugunsten eines inhaltlichen Zusammenhangs weicht der Übersetzer teilweise von der üblichen Satzstellung ab). Es ist zu begrüßen, daß keine gewaltsame Verdeutschung vorgenommen wurde, sondern die Fachausdrücke vielfach aus dem Englischen übernommen wurden (z. B. Konkatenation, AFL, Scattered-Kontext-Grammatik), da die überwiegende Zahl von Publikationen auf diesem Gebiet in englischer Sprache verfaßt ist. Der Wert des Buches wäre noch größer, wenn man die wesentlichen, seit dem Erscheinen des englischen Originals 1973 erschienenen Monographien als Ergänzung in das Literaturverzeichnis aufgenommen hätte.

Rostock J. DASSOW

Fritz, F.-J. / Huppert, B. / Willems, W., Stochastische Matrizen. Hochschultext. Berlin-Heidelberg-New York, Springer-Verlag 1979. IX, 193 S., DM 32,–, US $ 17,60. ISBN 3-540-09126-2

Im vorliegenden Buch werden endliche Matrizen mit nichtnegativen Elementen untersucht, deren Zeilensummen sämtlich gleich 1 sind. Da mit derartigen Matrizen das Übergangsverhalten einfacher stochastischer Prozesse beschrieben werden kann, spricht man von stochastischen Matrizen. Die Autoren verfolgen das Ziel, die in den ersten Semestern eines Hochschulstudiums vermittelten Anfangsgründe der Matrizentheorie unmittelbar auf Fragestellungen der Stochastik anzuwenden, ohne Vorkenntnisse in Wahrscheinlichkeitsrechnung vorauszusetzen. Dabei bleibt die Darstellung in Hinblick auf eine Theorie stochastischer Prozesse propädeutisch. Es geht hier nur um die stochastische Interpretation gewisser Klassen stochastischer Matrizen und ihrer Eigenwerte.

Somit lernt der Leser auf schöne Weise das Grenzverhalten homogener Markovketten in diskreter Zeit und endlichem Zustandsraum kennen. Unter den vielen Beispielen gefallen vor allem die Labyrinthprobleme.

Leipzig H.-J. GIRLICH

Chung, K. L., Elementare Wahrscheinlichkeitstheorie und stochastische Prozesse. Übers. aus d. Englischen. Hochschultext. Berlin-Heidelberg-New York, Springer-Verlag 1978. IX, 343 S., 36 Abb., DM 32,–. US $ 16.00.

Die Bedeutung dieses an Anfänger gerichteten Buches liegt nicht in der Stoffwahl — außer dem kanonischen Material, das für jede Einführung in die Wahrscheinlichkeitsrechnung obligatorisch ist, enthält das Buch etwas über die Grenzwertsätze, den POISSON-Prozeß und ein längeres Kapitel über MARKOWsche Ketten —, Hauptanliegen ist die gründliche Ausleuchtung der Begriffe und Methoden der elementaren Wahrscheinlichkeitstheorie. Dies geschieht mit einem Minimum an analytischem Aufwand. Eine Fülle von Beispielen, dazu fast dreihundert Übungsaufgaben mit Lösungen, eingestreute fast anekdotenhafte Bemerkungen, manchmal in der Diktion „onkelhafter Vertraulichkeit", wie der Autor das selbst nennt, geben ein ungewöhnlich farbiges Bild von besonderem Reiz. Die Hoffnung des Autors, „daß der Leser nicht nur etwas aus dem Buch lernt, sondern daß er dabei auch ein wenig Freude hat", ist bestimmt nicht vergeblich.

Berlin W. HEINRICH

Ibragimov, I. A. / Rozanov, Y. A., Gaussian Random Processes. Applications of Mathematics 9. Transl. from the Russian Original of 1970 by **Aries, A. B.,** Berlin-Heidelberg-New York, Springer-Verlag 1978. X, 275 S., 2 Abb., DM 49,60, US $ 27,30. ISBN 3-540-90302-X

This monography is a translation of the Russian edition "GAUSSovskie sluchainye protsessy" published in 1970 by Na kau. It treats GAUSSian stationary processes in an excellent way. The presented results are in many cases final solutions of the considered problems.

In the first chapters the authors are mainly clarifying the conditions for mutual absolute continuity of probability distributions of process segments. Further, effective formulas for densities of the equivalent distributions are considered.

In the following chapters classes of spectral measures are investigated which correspond to stationary processes (i.e. satisfying the strong mixing condition). Also the subclasses associated with mixing rate are described. This problem is closely connected with problems involving ergodic theory of GAUSSian dynamic systems and prediction theory of stationary processes. In the case of GAUSSian processes the consideration of regularity conditions reduces to a special approximation problem related to linear spectral theory. The book presents more or less completely the results on this field. The last chapter deals with statistical problems, (i. e. the problem of estimating the mean of a process). There are considered best unbiased estimates as well as least square estimates. The authors propose the class of "pseudobest" estimates and give explicit expressions, consistency conditions for asymptotic effectiveness and asymptotic formulas for the error correlation matrix. For reading and using this book it is necessary to be familar with the foundations of probability theory and stochastic processes.

This book is useful both as a source of reference for the more advanced mathematician and as a text for students.

Berlin E. PLATEN

Barlow, R. E. / Proschan, F., Statistische Theorie der Zuverlässigkeit. Berlin, Akademie-Verlag 1978. X, 250 S., M 48,–

In immer stärkerem Maße wird von technischen, aber auch von ökonomischen Systemen bei einer möglichst ständigen Betriebsbereitschaft und bei geringen Unterhaltungskosten eine hohe Zuverlässigkeit gefordert. Das gründet sich darauf, daß solche Systeme laufend größer und komplizierter werden und sich damit die Ausfallmöglichkeiten und deren Auswirkungen vergrößern. So hat in den letzten Jahren das Interesse an systematischen Untersuchungen von Zuverlässigkeitsproblemen beträchtlich zugenommen. Zur erfolgreichen Bewältigung der vielseitigen Aufgaben war und ist der Auf- und Ausbau einer geeigneten mathematischen Theorie eine wesentliche Grundlage. Die mathematische Zuverlässigkeitstheorie beschäftigt sich im Unterschied zu Zuverlässigkeitsfragen bei tragenden Konstruktionen, wo z. B. im Bauwesen und Maschinenbau Versagen infolge erstmaligen Überschreitens von kritischen Grenzen oder infolge Ermüdungsschadensakkumulation im Vordergrund stehen, hauptsächlich mit den vielfältigen Beziehungen zwischen einem System und seinen Komponenten. Sie hat u. a. die Aufgabe, mathematische Modelle für Zuverlässigkeitsprobleme aufzustellen und zu untersuchen und Kriterien zur Beurteilung der Zuverlässigkeit herauszuarbeiten. Das vorliegende Buch, eine Übersetzung der 1975 erschienenen englischen Ausgabe, ist als ein Standardwerk der mathematischen Zuverlässigkeitstheorie anzusehen und hilft, eine Lücke der deutschsprachigen Literatur auf diesem Gebiet zu schließen.

Im ersten Kapitel wird der zentrale Begriff des monotonen Systems eingeführt und erläutert. Das zweite Kapitel beschäftigt sich mit Beziehungen zwischen der Zuverlässigkeit eines Systems in Abhängigkeit von den Zuverlässigkeiten seiner Komponenten, wobei in einigen Fällen über die allgemein übliche Annahme der Unabhängigkeit der Komponenten hinausgegangen wird. Eine ausführlichere Behandlung der Abhängigkeiten zwischen den Komponenten mit Hilfe von mehrdimensionalen Verteilungen folgt im fünften Kapitel. Ausfallrate, Alterung und verschiedene Lebensdauerverteilungen, wie Exponential-, POISSON-, WEIBULL-, Gamma-, gestutzte Normalverteilung, sind Gegenstand des dritten Kapitels. In allgemeinerer Weise werden im vierten Kapitel Klassen von Verteilungen mit monotonen Ausfallraten behandelt und damit erzielte Schranken für die Zuverlässigkeit angeführt. Das sechste und siebente Kapitel sind der Untersuchung von verschiedenen Instandhaltungsstrategien gewidmet, insbesondere Fragen der Erneuerung, Verfügbarkeit, Reservierung und der optimalen Ersatzteilaufteilung. Im achten Kapitel sind Grenzverteilungen für die Lebensdauer spezieller Systemtypen angegeben. In einem Anhang wird die in Anwendungen durch Ingenieure häufig benutzte Methode des Fehlerbaumes erläutert.

Die einzelnen Paragraphen enthalten ausgewählte Übungsaufgaben, deren Lösung zum tieferen Verständnis des gebotenen Stoffes beiträgt. Die alle Kapitel abschließenden Bemerkungen und Literaturhinweise geben wichtige Ausblicke auf weiterführende Probleme und Spezialliteratur. Das Buch wendet sich an Lehrende und Lernende auf dem Gebiet der mathematischen Zuverlässigkeit und der Operationsforschung, für die es sich gewiß als große Hilfe erweisen wird.

Berlin H. FRIEDRICH

Bibby, J. / Toutenburg, H., Prediction and Improved Estimation. Chichester-New York-Brisbane-Toronto, John Wiley & Sons 1978. XIII, 188 S., US \$ 16.85, £ 7.95, ISBN 0-471-01656-X

Das vorliegende Werk ist eine lehrbuchartige Monographie zum Problem der optimalen Schätzung und der Vorhersage in linearen Regressionsmodellen. Es bietet in englischsprachiger, überarbeiteter Version den überwiegenden Teil des Materials aus TOUTENBURGS bekanntem Buch „Vorhersage in linearen Modellen" (1975, ZAMM **56** (1976), 445). In beiden Texten wird die statistische Theorie der linearen Modelle abgehandelt, wobei die vorliegende Darstellung indes spezifisch unter dem Aspekt der besten linearen Schätzung im Sinne der Minimierung des mittleren quadrierten Fehlers (mean squared error, MSE) des Schätzers aufgebaut ist. In diesem Sinne ist die „verbesserte Schätzung" zu verstehen: Unter allgemeinen Annahmen läßt sich der MSE reduzieren, wenn man von erwartungstreuen zu verzerrten Schätzern übergeht. Die konsequente Auswertung dieses Gesichtspunktes ermöglicht eine unkonventionelle und sehr anregende Präsentation des Standardmaterials wie auch zahlreicher allgemein weniger bekannter Entwicklungen auf dem Gebiet der Regressionsmodelle, so des Schätzers von STEIN, der RIDGE-Regression sowie anderer Techniken, durch bewußte Fehlspezifikation des Modells den MSE zu senken.

Das Inhaltsverzeichnis vermittelt einen Eindruck von der Fülle des behandelten Stoffes. Nach einem einleitenden motivierenden Kapitel („Basic ideas") wird das Konzept der „verbesserten Schätzung" („Improved methods in estimation") eingeführt und exemplifiziert. Es folgt die Darstellung der klassischen und der verbesserten Schätzung in linearen Modellen (Kapitel 3 und 4). Der zweite Hauptteil ist dem Prognoseproblem gewidmet (Kapitel 5–9). Zunächst wird hier die Theorie der besten Punktprognosen in großer Ausführlichkeit vorgeführt („Prediction in linear models", „Prediction with linear constraints"), wobei wiederum der Zugang unter dem Aspekt des MSE so interessante Fragen wie die Prognose aus fehlspezifizierten Modellen zu diskutieren gestattet („Prediction and model choice"). Sodann wird die Theorie der Prognosebereiche entwickelt („Prediction intervals") und die Prognosetheorie auf multivariate Regressionen („Prediction in econometric models") angewendet. Ein Kapitel über Minimax-Schätzer in linearen Modellen, das dem ursprünglichen Text von TOUTENBURG fehlt, ergänzt den Haupttext. Zwei Anhänge über Matrixalgebra und Matrixdifferentiation erleichtern das Textverständnis.

Insgesamt bietet das neue Werk eine gelungene Überarbeitung und Erweiterung des Buches von TOUTENBURG, wobei die Präzision des ursprünglichen Textes erhalten geblieben ist, die Lesbarkeit jedoch durch eine noch ausgefeiltere didaktische Präsentation und durch Anreicherung mit motivierenden Anwendungen erhöht worden ist, so daß das Werk einem allgemeineren Leserkreis als Lehrbuch der Regressionstheorie dienen kann.

Bonn P. SCHÖNFELD

Toutenburg, H. / Rödel, E., Mathematisch-statistische Methoden in der Ökonomie. Berlin, Akademie-Verlag 1978. 192 S., M 28,–.

Das vorliegende Buch gibt auf knapp 200 Seiten einen Überblick über wichtige und praktikable Methoden in der Ökonometrie (vorwiegend Schätz- und Vorhersagemethoden), wobei Hauptaugenmerk auf Einfachheit der Verfahren und möglichst vollständige Ausnutzung vorhandener Zusatzinformationen gelegt wird. — Inhalt: 1. Einleitung; 2. Klassische lineare Regression; 3. Das verallgemeinerte Regressionsmodell; 4. Schätzung unter linearen Restriktionen; 5. Vorhersage und Modellwahl; 6. Probleme der Schätzung und Vorhersage in linearen stochastischen Modellen; 7. Zeitreihenanalyse; 8. Simultane Gleichungssysteme — Das allgemeine ökonometrische Modell; Anhang A (vorwiegend Matrix-Algebra; Matrix-Differentiation); Anhang B (stationäre stochastische Prozesse). — Wenn innerhalb von 3 Jahren im gleichen Verlag vom gleichen Autor (diesmal mit Co-Autor) ein Buch gleicher Aufmachung zu annähernd gleichem Problemkreis erscheint, so ist dies ein Zeichen für die Aktualität des Stoffes und das Können der Autoren, diesen verständlich zu vermitteln.

Die ersten sechs Abschnitte des vorliegenden Buches überdecken inhaltlich weitgehend entsprechende Abschnitte aus oben erwähntem Buch TOUTENBURG, Vorhersage in linearen Modellen, Akademie-Verlag 1975 (Neu aufgenommen wurden vor allem Minimax-lineare Schätzungen, zu denen der erste Autor in letzter Zeit einige Originalarbeiten veröffentlich hat). Die vorliegende Fassung hat jedoch gegenüber der alten deutlich gewonnen, da einige mehr nur theoretische Betrachtungen weggelassen wurden, die Spezifik der einzelnen vorgestellten Methoden deutlicher herausgearbeitet und die Tragfähigkeit der Verfahren u. a. in Teilmodellen der Volkswirtschaft der DDR demonstriert wurde. Die letzten beiden Abschnitte, verfaßt vom zweiten Autor, geben eine gelungene Einführung in die Zeitreihenanalyse (vor allem Schätzen der Trendfunktion und der Spektraldichte) und in die simultanen Gleichungssysteme. Das Studium des nützlichen Buches erfordert einige Grundkenntnisse in Wahrscheinlichkeitstheorie und Statistik, sowie in Matrixalgebra.

Einige kritische Bemerkungen: Gerade für Belange der Ökonomie hätte man sich eine etwas ausführlichere Diskussion der oft verwendeten Begriffe Verlustfunktion und Risiko gewünscht (zumal beim ersten Auftreten des Begriffes Risiko in Formel (2.4.8) eine insofern unglückliche Verlustfunktion gewählt wurde, als ihre Werte vom Experimentator (genauer: von der Festlegung der Matrix X) abhängen. — Bei den Methoden zum Ausnutzen von Zusatzinformationen in Form von Parameterrestriktionen fehlt evtl. eine Einschätzung und Wertung der z. T. sehr in Mode gekommenen BAYESschen Methoden. — Im Abschnitt Zeitreihenanalyse hätten vielleicht auch andere Entwicklungen zitiert werden sollen, vor allem die in den letzten Jahren stark ausgebauten ARIMA-Modelle (z. B. in BOX/JENKINS, Time Series, Forecasting and Control, Holden-Day 1970).

Freiberg W. NÄTHER

Cigler, J. / Reichel, H.-C., Topologie. Eine Grundvorlesung. B. I. Hochschultaschenbücher Band 121. Mannheim-Wien-Zürich, B. I.-Wissenschaftsverlag 1978. XIII, 244 S., ISBN 3-411-00121-6

Dieses Buch ist als Einführung vor allem in die mengentheoretische Topologie, d. h. die Theorie der topologischen Räume und der Konvergenz gedacht. Es umfaßt — außer der Einleitung, die die für die Topologie wichtigen Grundlagen aus der Mengenlehre bringt, darunter den gesamten Themenkreis, der sich um das Auswahlaxiom gruppiert — die Kapitel: Topologische Räume; Metrische Räume; Vollständig reguläre Räume, Pseudometriken (Uniforme Räume I); Kompakte Räume; Zusammenhängende Räume; Homotopie und Fragen der Topologie des R^n sowie Uniforme Räume (2. Teil). Schon daraus ersieht man, daß es sich um eine sehr aktuelle Darstellung des Stoffes handelt (übrigens wird weitgehend mit MOORE-SMITH-Folgen gearbeitet und nicht mit Filtern, wie etwa in dem bekannten Buch von KOWALSKI). Beim Lesen des recht anregend geschriebenen Buches bestätigt sich dieser Eindruck vollauf. Es enthält darüber hinaus eine Fülle von Beispielen und Aufgaben, die größtenteils in weitere, im Buch nicht behandelte Gebiete führen, in die man sich dank des guten Literaturverzeichnisses bei Bedarf hineinlesen kann. Einem Leser, der das Buch aufmerksam liest und der die Aufgaben weitgehend löst, dürfte es nicht schwer fallen, Anschluß an die Forschung zu bekommen. Es ist ‚reinen' wie ‚angewandten' Mathematikern wärmstens zu empfehlen, denn auch für letztere ist Topologie wesentliche Grundlage geworden.

Berlin H. WEINERT

Nowacki, W. / Sneddon, I. N. (eds.), Thermomechanics in Solids. International Centre for Mechanical Sciences. CISM Courses and Lectures 223. Wien-New York, Springer-Verlag 1977. IV, 87 S., ÖS 152,– DM 22,–. ISBN 3-211-81343-8

Namhafte Vertreter der Thermomechanik stellen in diesem Heft 4 Beiträge vor: 1. I. N. SNEDDON, Coupled problems in the linear theory of thermoelasticity; 2. W. NOWACKI, Thermal stresses in anisotropic bodies; 3. H. PARKUS, Magneto- and electrothermoelasticity; 4. C. WOCNIAK, On the thermoelasticity of nonlinear discrete and continous constrained systems.

Mit dieser Problematik führten die Autoren im CISM in Udine im Jahre 1974 die RANKINE Session ein. Es handelt sich um knappe Übersichten zu den angeführten Themen. Stets wird ausschließlich die Theorie sauber hergeleitet, ohne auf spezielle Beispiele einzugehen. Wegen der vielen Literaturstellen ist es allen, die sich mit dieser Thematik beschäftigen, aber auch denjenigen, die sich erst einarbeiten wollen, sehr zu empfehlen.

Dresden H. GÖLDNER

Hutter, K. / van de Ven, A. A. F., Field Matter Interactions in Thermoelastic Solids. A Unification of Existing Theories of Electro-Magneto-Mechanical Interactions. Lecture Notes in Physics 88. Berlin-Heidelberg-New York, Springer-Verlag 1978. VIII, 231 S., DM 25,–. US $ 13,80. ISBN 3-540-09105-X

In den letzten Jahrzehnten sind immer kompliziertere Modelle für thermoelastische Körper entstanden, indem die verschiedensten physikalischen Einflüsse und Wechselwirkungen berücksichtigt werden. Viele Theorien sind schwer vergleichbar, nicht zuletzt deshalb, weil unter demselben Namen oft Verschiedenes verstanden wird.

Im vorliegenden Buch werden fünf verschiedene bekannte Beschreibungen für deformierbare polarisierbare und magnetisierbare Kontinua dargelegt, wobei jedes Modell auf seine Weise die Wechselwirkungen zwischen dem elektromagnetischen Feld und dem thermoelastischen Körper beschreibt. Die Modelle beinhalten 34 Feldgleichungen (MAXWELLsche Gleichungen (7), Bilanzgesetze für Masse, Impuls und Energie (5), Beziehungen für Entropie, zwei elektromagnetische Feldvektoren und den Spannungstensor (16), Beziehungen für den elektrischen Strom und den Energieflußvektor (6)) für 34 Unbekannte (4 elektromagnetische Feldvektoren (16), elektrischer Strom und elektrische Ladungsdichte (4), Massendichte (1), Bewegung (3), Spannung (9), Temperatur (1), Energiefluß (3), Entropie (1)). Es wird bewiesen, daß alle fünf Modelle (2-Dipol-Modell mit unsymmetrischem und symmetrischem Spannungstensor, MAXWELL-MINKOWSKI Formulierung, statistische Formulierung, Lorentz-Formulierung) im Rahmen der zugrunde gelegten nichtrelativistischen Approximation im gewissen Sinne äquivalent sind.

Die Gleichungen der fünf Modelle werden auch in der für die Beschreibung der Deformationen von Festkörpern günstigeren materiellen oder LAGRANGEschen Formulierung hergeleitet, die auch den Ausgangspunkt für einige in der Arbeit durchgeführte Linearisierungen bildet. Der Vergleich der verschiedenen elektromechanischen Wechselwirkungstheorien ist durch eine eingehende Beleuchtung von verschiedenen Seiten gut gelungen und trägt zur Herausbildung klarer Modelle bei, die für eine mathematische Durchdringung wichtige Voraussetzung sind.

Leipzig L. JENTSCH

Dixon, W. G., Special Relativity. The Foundation of Macroscopic Physics. Cambridge-London-New York-Melbourne, Cambridge University Press 1978. X, 261 S., £ 14.50.

Das 1978 in erster Auflage erschienene Werk erweitert die bereits umfangreiche Literatur zur Speziellen Relativitätstheorie um ein weiteres Buch, in dem die einheitliche Behandlung von Mechanik, Elektromagnetik und Thermodynamik im Vordergrund steht. Um dem Leser einen leichteren Zugang zu verschaffen, werden aus didaktischen Gründen die Newtonsche und die Relativistische Physik streckenweise parallel dargelegt. Den Schwerpunkt des Buches, das sich durchweg des eleganten Indexkalküls bedient, bildet die speziell-relativistische Elektrodynamik polarisierbarer Fluids. Dabei kommt die Thermodynamik voll zur Geltung: Energiesatz, Entropiesatz, lokales und globales Gleichgewicht, GIBBSsche Fundamentalform, lineare Materialgleichungen, Entropieproduktion, elektromagnetische Dispersion. Diese Stichworte charakterisieren die zweite Hälfte des Buches, das für Studenten der Physik und angewandten Mathematik ab 2. Studienjahr gedacht ist, in der Regel aber doch wohl erst von Studenten höherer Studienjahre richtig verstanden werden kann.

Jena E. SCHMUTZER

In der DDR beziehbar im Rahmen des Kontinents.

EINGEGANGENE BÜCHER

Bei der Schriftleitung sind folgende Bücher eingegangen (ausführliche Besprechung bleibt vorbehalten).

Nashed, M. Z. (Ed.), Functional Analysis Methods in Numerical Analysis. Special Session, American Mathematical Society, St. Louis, Missouri 1977. Lecture Notes in Mathematics 701. Berlin-Heidelberg-New York, Springer-Verlag 1979. VII, 333 S., 1 Abb., 4 Tab., DM 32,–, US $ 17,60. ISBN 3-540-09110-6

Grosswald, E., Bessel Polynomials. Lecture Notes in Mathematics 698. Berlin-Heidelberg-New York, Springer-Verlag 1978. XIV, 182 S., 11 Abb., 1 Tab., DM 21,50, US $ 11,90. ISBN 3-540-09104-1.

Feinsilver, P. J., Special Functions, Probability Semigroups, and Hamiltonian Flows. Lecture Notes in Mathematics 696. Berlin-Heidelberg-New York, Springer-Verlag 1978. VI, 112 S., DM 18,–, US $ 9,90. ISBN 3-540-09100-9.

Wonham, W. M., Linear Multivariable Control: A Geometric Approach. Second Edition. Applications of Mathematics 10. Berlin-Heidelberg-New York, Springer-Verlag 1979. XV, 326 S., 27 Abb., DM 62,–, US $ 34,10. ISBN 3-540-90354-2.

Fábera, J. (Ed.), Equadiff IV. Proceedings, Prague, August 22–26, 1977. Lecture Notes in Mathematics 703. Berlin-Heidelberg-New York, Springer-Verlag 1979. XIX, 441 S., 9 Abb., 2 Tab., DM 42,–, US $ 23,40. ISBN 3-540-09116-5.

Weil, André, Œuvres Scientifiques – Collected Papers. Volume I (1926–1951), Volume II (1951–1964), Volume III (1964–1978). Berlin-Heidelberg-New York, Springer-Verlag 1979. Vol. I: XVIII, 574 S.; Vol. II: XII, 561 S.; Vol. III: XII, 465 S. Insgesamt DM 160,–, US $ 88,00, ISBN 3-540-90330-5.

Wilhelmsson, H. (Ed.), Solitons in Physics. Physica Scripta, Vol. 20, 3/4 September–October, 1979, S. 289–562. The Royal Swedish Academy of Sciences, Publications Department, Stockholm, Sw.kr. 250. ISBN 0031-8949.

Bain, L. J., Statistical Analysis of Reliability and Life-Testing Models. Theory and Methods. Statistics: Textbooks and Monographs, Vol. 24. New York and Basel, Marcel Dekker, Inc. 1978. XII, 450 S., 62 Tab., SFr. 80. 00. ISBN 0-8247-6665-2.

Richter, G., Kategorielle Algebra. Studien zur Algebra und ihre Anwendungen, Band 3. Berlin, Akademie-Verlag 1979. 299 S., 58,– M. Best.-Nr. 762 3774 (2174/3).

Färber, G., Prozeßrechentechnik. Allgemeines, Hardware und Software, Planungshinweise. Nachrichtentechnik 5. Berlin-Heidelberg-New York, Springer-Verlag 1979. X, 208 S., 98 Abb., 5 Tab., DM 48,–, US $ 26.40. ISBN 3-540-09263-3.

Ahlswede, R. / **Wegener, I.,** Suchprobleme. Teubner Studienbücher Mathematik. Stuttgart, B. G. Teubner 1979. 328 S., 22 Abb., DM 28,80. ISBN 3-519-02058.

Franksen, O. I. / **Falster, P.** / **Evans, F. J.,** Qualitative Aspects of Large Scale Systems. Developing Design Rules Using APL. Lectures Notes in Control and Information Sciences 17. Berlin-Heidelberg-New York, Springer-Verlag 1979. XII, 119 S., DM 18,–, US $ 9.90. ISBN 3-540-09609-4.

Renken, J., Potentialtheoretische und experimentelle Untersuchung der dreidimensionalen Unterschallströmung in geraden Verdichtergittern. DFVLR Deutsche Forschungs- und Versuchsanstalt für Luft- und Raumfahrt DFVLR-FB 79-09. Köln, Wissenschaftliches Berichtswesen der DFVLR 1979. 182 S., 138 Abb., DM 33,30.

Keller, H. P., Über das J-Integral und andere Auswerteverfahren der Fließbruchmechanik sowie über die Gültigkeitsgrenzen der linear-elastischen Bruchmechanik. DFVLR Deutsche Forschungs- und Versuchsanstalt für Luft- und Raumfahrt DFVLR-FB 79-03. Köln, Wissenschaftliches Berichtswesen der DFVLR 1979. 168 S., 70 Abb., 6 Tab., DM 24,90.

Nowak, P., Bildverbesserung an multispektralen Scanneraufnahmen mit Hilfe digitaler Filterverfahren. DFVLR Deutsche Forschungs- und Versuchsanstalt für Luft- und Raumfahrt DFVLR-FB 79-11. Köln, Wissenschaftliches Berichtswesen der DFVLR 1979. 158 S., 90 Abb., 13 Tab. DM 28,–.

Walther, H., Anwendungen der Graphentheorie. Braunschweig-Wiesbaden, Friedr. Vieweg & Sohn 1979. 239 S., 102 Abb., DM 39,80. ISBN 3-528-08418-9.

NACHRICHTEN

Future IUTAM Symposia

Whereas participation in the International Congresses of Theoretical and Applied Mechanics is unrestricted, all IUTAM Symposia are reserved for invited participants.

Those wishing to participate actively in an IUTAM Symposium are therefore advised to contact the Chairman of the Scientific Committee under the address given below far in advance of the meeting.

1980

1. IUTAM Symposium on Physical Nonlinearities in Structural Analysis

Place: Senlis, France. Date: 27–30 May 1980
Chairmen: Professor J. Lemaitre, Laboratoire de Mécanique, Ecole Normale Superieure de l'Enseignement Technique, F-94230 CACHAN, France
Professor J. Hult, Chalmers University, of Technology, S-40220 GOTHENBURG, Sweden

2. IUTAM Symposium on Threedimensional Constitutive Relations

Place: Palaiseau, France. Date: 2–5 June 1980
Chairman: Professor J. Mandel, Laboratoire de Mécanique des Solides, Ecole Polytechnique, F-91128 PALAISEAU, Cedex, France

3. IUTAM Symposium on Finite Elasticity Theory

Place: Bethlehem, Pennsylvania, USA. Date: 11–15 August 1980
Chairman: Professor R. S. Rivlin, Lehigh University, Center for the Application of Mathematics, 203 E. Packer Ave., BETHLEHEM, PA 18015, USA

4. Third IUTAM Symposium on Creep in Structures

Place: Leicester, UK. Date: 8–12 September 1980
Chairman: Professor A. R. S. Ponter, Department of Engineering, The University, LEICESTER LEI 7RH, UK

5. ICHMT/IUTAM Symposium on Heat and Mass Transfer and the Structure of Turbulence

Place: Dubrovnik, Yugoslavia. Date: Not yet decided
Chairman: Professor Z. Zaric, International Centre for Heat and Mass Transfer, P. O. Box 522, 11001 BEOGRAD, Yugoslavia

1981

1. IUTAM/IUPAC Symposium on Interaction of Particles in Colloidal Dispersions

Place: Canberra, Australia. Date: 16–21 March 1981 (tentative)

Chairman: Professor B. W. Ninham, Research School of Physical Sciences, The Australian National University, CANBERRA, A. C. T. 2600, Australia

2. CISM/IUTAM Symposium on Crack Formation and Propagation

Place: Warsaw, Poland. Date: 23–27 March 1981
Chairman: Professor W. Olszak, CISM, Palazzo del Torso, Piazza Garibaldi, 18, I-33100 UDINE, Italy

ICF/IUTAM Symposium on the Foundations of Fracture Mechanics
Cancelled

3. IUTAM Symposium on the Mechanics and Physics of Bubbles in Fluids

Place: Pasadena, California, USA. Date: 15–19 June 1981
Chairmen: Professor M. S. Plesset, California Institute of Technology, Engineering Science Department, PASADENA, CA 91125, USA
Professor L. van Wijngaarden, Technische Hogeschool Twente, Postbus 217. ENSCHEDE, the Netherlands

4. IUTAM Symposium on Mechanics of Composite Materials

Place: Blacksburg, Virginia, USA (tentative). Date: June 1981 (tentative)
Chairman: Not yet appointed

5. IUTAM/IUGG Symposium on Intense Vortices

Place: Reading, UK. Date: July 1981 (tentative)
Chairman: Not yet appointed

6. Second IUTAM Symposium on Stability in the Mechanics of Continua

Place: Wuppertal, BRD. Date: 24–28 August 1981
Chairman: Professor F. Schroeder, Bergische Universität – Gesamthochschule Wuppertal, D-5600 WUPPERTAL, BRD

7. IUTAM Symposium on High Temperature Gas Dynamics

Place: Prague, Czechoslovakia. Date: Not yet decided
Chairman: Dr. M. Pichal, Insitute of Thermomechanics, Czechoslovak Academy of Sciences, Puškinovo nám. 9, PRAHA 6, Czechoslovakia

1982

1. IUTAM Symposium on Unsteady Turbulent Shear Flows

Place: Toulouse, France. Date: Not yet decided
Chairman: Dr. R. Michel, Centre d'etudes et de Recherche de Toulouse, BP No 4025, F-31055 TOULOUSE, Cedex, France

2. IUTAM Symposium on Threedimensional Turbulent Boundary Layers

Place: Berlin (West) or Aachen, BRD. Date: Not yet decided
Chairmen: Professor H. H. Fernholz, Hermann-Föttinger-Institut für Thermo- und Fluiddynamik, Technische Universität, Straße des 17. Juni, 135, D-1000 BERLIN 12
Professor E. Krause, Aerodynamisches Institut, Technische Hochschule, AACHEN, BRD

3. ISIMM/IUTAM Symposium on Modern Development of Analytical Mechanics

(ISIMM = International Society for the Interaction of Mechanics and Mathematics)
Place: Not yet decided. Date: Not yet decided
Chairman: Professor D. Galletto, Istituto di Meccanica Razionale, Università di Torino, TORINO, Italy

International centre for mechanical sciences

"Structural Mechanics Problems in Offshore Engineering"
a short intensive course for engineers and researchers, Udine, September 15–19, 1980.

The 1980 course, the second in the series, will treat in a systematic way the peculiar aspects of the structural analysis and design of offshore constructions (platforms, risers, pipelines), with emphasis on steel structures. Methods for the computerized approach, the static, dynamic and buckling analysis of overall structures or structural components, safety concepts and the main design criteria will be given an up-to-date presentation and critical survey.

The lectures, (about 35 hours) will be delivered by C. Brebbia (University of California), S. J. Britvec (University of Colorado), E. Gnone (Tecnomare, Milan), C. Kirk (Cranfield Institute of Technology), J. Leonard (University of Oregon), R. Matteelli (Snamprogetti, Milan), T. Moan (Norwegian Institute of Technology), A. C. Palmer (University of Manchester).

The registration fee amounts to 250 US $. Information and application forms can be requested at:

C. I. S. M., Piazza Garibaldi 18, **33100 Udine** (Italy), tel (0432) – 294989 or 22523

Forschungsarbeiten zur Strömungsturbulenz

Zwei- und dreidimensionale Austauschvorgänge

Herausgegeben von Manfred Hoffmeister

(Schriftenreihe des Zentralinstituts für Mathematik und Mechanik der Akademie der Wissenschaften der DDR)

1979. 189 Seiten — 165 Abbildungen — 11 Tabellen — gr. 8° — 48,— M
Bestell-Nr. 7626908 (2069/28)

Der Band enthält Arbeiten zum turbulenten Wärmetransport in Abhängigkeit von der Grobstruktur der Turbulenz sowie Betrachtungen zur ebenen turbulenten Parallelströmung bei verschwindender Wandschubspannung. Durchgeführte Untersuchungen der reibungsbehafteten Umströmung an Schiffsdoppelmodellen im Windkanal werden beschrieben und Spektralanalysen turbulenter Geschwindigkeitsschwankungen in einem Modellrührwerk aufgezeichnet.

Bestellungen durch eine Buchhandlung erbeten

Akademie-Verlag
DDR-1080 Berlin, Leipziger Str. 3—4

W. M. GLUSCHKOW, G. J. ZEITLIN und J. L. JUSTSCHENKO

Algebra, Sprachen, Programmierung

(Übersetzung aus dem Russischen)

In deutscher Sprache herausgegeben von Rolf Lindner und Hansgeorg Meißner

1980. Etwa 336 Seiten — 34 Abbildungen — 15 Tabellen — gr. 8° — Lederin etwa 66,— M
Bestell-Nr. 762 2990 (6354)

Gegenwärtig kreuzen sich in der Computer-Theorie und -Praxis mehrere Entwicklungslinien. Einerseits wird eine mathematisch fundierte Bereitstellung leistungsfähiger Software zur weiteren Verbesserung der Nutzfreundlichkeit großer Rechner verlangt, andererseits erhebt sich die mit der Entwicklung der Mikroelektronik (wieder) auf den Plan gerufene Forderung nach gut handhabbaren mathematischen Methoden der maschinenorientierten (Mikro)-Programmierung. Daraus erklären sich viele Versuche, mit typisch algebraischen Mitteln zur Lösung der Probleme beizutragen. Hier ordnen sich auch die von den Autoren entwickelten angewandten algebraischen Methoden ein.

Bestellungen durch eine Buchhandlung erbeten

Akademie-Verlag
DDR-1080 Berlin, Leipziger Str. 3—4

FSC
www.fsc.org
MIX
Papier | Fördert gute Waldnutzung
FSC® C083411

Zeitfracht Medien GmbH
Ferdinand-Jühlke-Straße 7
99095 Erfurt, Deutschland
produktsicherheit@kolibri360.de